Fixed Points

ALGORITHMS and APPLICATIONS

Academic Press Rapid Manuscript Reproduction

Proceedings of a Conference on Computing Fixed Points with Applications held in the Department of Mathematical Sciences at Clemson University, Clemson, South Carolina, June 26-28, 1976.

International Conference on Computing
Fixed Points with Applications, 1st,
Clemson University, 1974.

Fixed Points

ALGORITHMS and APPLICATIONS

EDITED BY

STEPAN KARAMARDIAN

Graduate School of Administration
and Department of Mathematics
University of California, Irvine
Irvine, California

In collaboration with

C. B. Garcia

Graduate School of Business
University of Chicago
Chicago, Illinois

ACADEMIC PRESS, INC.

New York   San Francisco   London   1977
A Subsidiary of Harcourt Brace Jovanovich, Publishers

ACADEMIC PRESS, INC.
111 Fifth Avenue, New York, New York 10003

United Kingdom Edition published by
ACADEMIC PRESS, INC. (LONDON) LTD.
24/28 Oval Road, London NW1

Library of Congress Cataloging in Publication Data

International Conference on Computing Fixed
    Points with Applications, 1st, Clemson Univer-
    sity, 1974.
    Fixed points.

    "Sponsored by the Office of Naval Research and
the Office of the Army Research Center."
    Bibliography: p.
    Includes index.
    1.   Fixed point theory—Congresses.   2.   Equa-
tions, Simultaneons—Congresses.   3.   Economics,
Mathematical—Congresses.   4.   Programming (Mathe-
matics)—Congresses.   I.   Karamardian, Stepan.
II.   Garcia, C. B.   III.   United States.   Office
of Naval Research.   IV.   United States.   Army
Research Office.   V.   Title.
QA329.9.I57   1974   515   76-30678
ISBN 0–12–398050–X

Contents

Contributors

E. L. ALLGOWER, Department of Mathematics, Colorado State University, Fort Collins, Colorado 80521

CHRISTOPHER BOWMAN, Department of Mathematics, University of California, Irvine, California 92717

RICHARD W. COTTLE, Department of Operations Research, Stanford University, Stanford, California 94305

PHILLIPPE J. DESCHAMPS, CORE DeCroylaan 54, 3030 Haverlee, Belgium

B. CURTIS EAVES, Department of Operations Research, Stanford University, Stanford, California 94305

M. L. FISHER, Graduate School of Business, University of Chicago, Chicago, Illinois 60637

JOHN FREIDENFELDS, Bell Labs, Room 2A, 206 Whippany, New Jersey 07981

C. B. GARCIA, Graduate School of Business, University of Chicago, Chicago, Illinois 60637

F. J. GOULD, Graduate School of Business, University of Chicago, Chicago, Illinois 60637

C. W. GROETSCH, Department of Mathematical Sciences, University of Cincinnati, Cincinnati, Ohio 45221

TERJE HANSEN, Norwegian School of Economics and Business Administration, 5000 Bergen, Norway

STEPAN KARAMARDIAN, Graduate School of Administration, University of California, Irvine, California 92717

R. B. KELLOGG, University of Maryland, Institute for Fluid Dynamics and Applied Mathematics, College Park, Maryland 20742

HAROLD W. KUHN, Department of Mathematics, Princeton University, Princeton, New Jersey 08540

T. Y. LI, University of Maryland, Institute of Fluid Dynamics and Applied Mathematics, College Park, Maryland 20742

JAMES MACKINNON, Department of Economics, Dickinson Hall, Princeton

R. SAIGAL, Bell Labs, Holmdel, New Jersey 07733

HISAKAZU NISHINO, Project on Efficiency and Decision Making, Harvard University, Cambridge, Massachusetts 02138

B. E. RHOADES, Department of Mathematics, Indiana University, Swain Hall-East, Bloomington, Indiana 47401

ALVIN E. ROTH, Department of Business Administration, University of Illinois, Urbana, Illinois 61801

R. SAIGAL, Bell Labs, Holmdel, New Jersey 07733

HERBERT E. SCHARF, Department of Economics, Yale University, New Haven, Connecticut

JOHN B. SHOVEN, Department of Economics, Stanford University, Stanford, California 94305

MICHAEL J. TODD, Department of Operations Research, Cornell University, Ithaca, New York 14850

J. W. TOLLE, Department of Mathematics, University of North Carolina, Chapel Hill, North Carolina

JOHN WHALLEY, The London School of Economics and Political Science, Houghton Street, London, England WC2A2AE

RICHARD WILMUTH, International Business Machines Corporation, Monterey and Cottle Roads, San Jose, California 95114

J. YORKE, University of Maryland, Institute of Fluid Dynamics and Applied Mathematics, College Park, Maryland 20742

Preface

Since the appearance of Brouwer fixed point theorem in 1912 and its subsequent generalizations, fixed point theorems provided powerful tools in demonstrating the existence of solutions to a large variety of problems in applied mathematics. However, from the computational standpoint, their usefulness was limited. Up to 1967 all computational methods used for computing an approximate fixed point for a given map were based on iterative procedures that required additional restrictions on the map to guarantee convergence.

In 1967 H. Scarf developed a finite algorithm for approximating a fixed point to a continuous map from a simplex into itself. Scarf's algorithm is based on first subdividing the simplex into finite subsets called primitive sets and then utilizing Lemke's complementarity pivoting procedure. This algorithm provided the first constructive proof to Brouwer's fixed point theorem.

Scarf's work stimulated considerable interest. During the following few years several important refinements and extensions to his algorithm were developed. Among those are the works of H. Kuhn, B. C. Eaves, and O. Merrill.

These algorithms were applied to a number of test and real problems, with reasonable degrees of efficiency. During this period there was also considerable interest in a related unifying model, the so-called "complementarity problem," the problem of finding a nonnegative vector whose image under a given map is also nonnegative, and such that the two vectors are orthogonal. In fact, the fixed point problem over the nonnegative orthant of a finite-dimensional Euclidean space is equivalent to a complementarity problem.

In early 1974 several researchers expressed an interest in holding a conference to bring together those who were active in the fields of fixed point algorithms, the complementarity problem, and those who were involved in their applications to economics and other problems.

The first International Conference on Computing Fixed Points with Applications was held in the Department of Mathematical Sciences at Clemson University, Clemson, South Carolina, June 26-28, 1974.

The Conference was sponsored by the Office of Naval Research and the Office of the Army Research Center. The participants included mathematicians and economists from several European countries, Japan, and the United States.

Nine one hour invited addresses and twelve one half hour contributed papers were presented during the Conference. Each presentation was followed by a half hour discussion period. All papers were refereed, edited, and finally approved by the authors before their publication in this volume.

Professor Herbert Scarf who attended the Conference and participated actively in the discussions was kind enough to write a very illuminating introduction to the proceedings.

The organizing committee consisted of J. Kenelly, C. J. Ancoin, C. B. Garcia, and S. Karamardian. Mrs. M. Hinton was most helpful in her role as secretary of the conference. Ms. Leslie Cobb typed the manuscripts, Ms. Gini Nordyke drew the technical figures; and Ms. Lynn Mayeda did the final proof-reading. We extend our warmest appreciation to these dedicated persons and organizations.

Introduction

Herbert E. Scarf

I would like to take this opportunity to give a personal view of some of the major developments in the field of
fixed point computations during the last half dozen years.
As the present volume clearly indicates, the field has prog-
ressed substantially during this brief period. My selection
of topics will necessarily omit many contributions of con-
tinuing importance, which a more leisurely discussion would
include.

It may be difficult to recapture the view which many of
us had some ten years ago -- that the problems of mathematical programming fall naturally into two quite distinct cate-
gories in terms of whether or not they can be solved numerically. On the one hand, there were powerful techniques for
solving the linear and non-linear programming problems arising when the production side of the economy is studied in
isolation. But on the other hand as soon as a number of independent agents were introduced into the problem -- either
as players in an n-person game or consumers in a model of
economic equilibrium -- we were forced to be content with
theorems advising us of the existence of a solution but with
no indication whatsoever of a constructive method for its
determination.

By the late 1960's -- the date which I would like to
take as the point of departure for the present discussion --
this difficulty had, of course, been overcome. A variety of
novel computational techniques had been developed for the
approximation of the fixed points of a continuous mapping or
correspondence. A substantial number of numerical examples

had been successfully attempted and it had become clear that
the efficiency of these algorithms was sufficiently high so
as to justify their application to realistic problems of moderate size. Moreover, these algorithms were sufficiently
flexible, so that a given problem might be solved by one of
several methods -- selected so as to exploit the features of
the particular problem in question. A survey of the state of
the art at this moment of time, together with a number of
specific applications may be found in the monograph, <u>The Computation of Economic Equilibria</u>, written in collaboration
with Terje Hansen.

In order to introduce the discussion of recent developments, it is useful to remark on one major drawback of the
early techniques: they typically required that the computational procedures be initiated on the boundary of the underlying simplex on which the mapping was defined -- in some instances at a vertex of the simplex -- and that a rough measure of accuracy be assigned in advance. If, after the com-
pletion of the computation, the accuracy were judged to be
insufficient, the only available recourse was to perform the
entire computation again with a finer grid size. The results
of the earlier computation, which provided a rough indication
of the location of the answer, were completely discarded on
the grounds that they could not serve as the starting point
for a subsequent attempt at higher precision.

The ability to obtain high precision -- say seven or
eight significant digits -- with fixed point methods may be
of considerable importance even if the underlying data of
the problem lack a similar degree of accuracy. In order to
estimate the consequences of a particular change in economic
policy -- for example, an increase in tariffs -- a general

equilibrium model will be solved numerically both before and after the imposition of that policy. But then in order to assert that the policy will lead to a 5% increase in the price of a certain commodity or a 10% decrease in its level of production, these prices and production levels must be calculated with sufficiently high accuracy to reflect the difference in their values.

There are, of course, well established numerical techniques, such as Newton's method, which converge locally to the solution of a system of non-linear equations. In the early programs, I frequently tacked on, after the fixed point approximation, a relatively crude variant of Newton's method which was written specifically for the problem at issue. The results were sufficiently satisfactory so that I know of at least one user who discards fixed point methods completely and is content with guessing an answer which is subsequently refined by Newton's method.

The major drawback of Newton's method -- aside from its apparent lack of harmony with fixed point techniques -- is that a specific program is required for any basic variation in approach to the problem being analyzed. A general equi-librium model, to take one example, may be solved in three of four quite distinct ways depending on its special structure; ideally one would like a method for obtaining high accuracy which is independent of the specific technique selected.

One of the major advances of recent years -- introduced by Curtis Eaves -- may be viewed as a technique which permits a continued improvement in accuracy without recourse to Newton's method, though its ramifications are considerably greater than this somewaht technical justification would suggest. As distinct from earlier authors, Eaves does not work

with a simplicial subdivision of the simplex on which the
mapping is defined, but rather with a subdivision of the cylinder formed by taking the product of this simplex with a finite interval. The mapping whose fixed point we wish to determine is placed on one end of the cylinder. A trivial mapping of the simplex into itself, whose fixed point is unique
and can be placed in an arbitrary location, appears on the
opposite end. The two mappings are then joined by a piecewise linear homotopy throughout the cylinder. Given this
setting, the earlier methods of simplicial pivoting can be
extended so as to construct an algorithm which begins at the
pre-assigned location on the end bearing the trivial mapping,
and which terminates on the opposite end with an approximate
fixed point of the mapping in question.

At the cost of one additional dimension, Eaves' methods
permit us to initiate the computation at an arbitrary point
and to continue without starting the procedure again, until
the desired degree of accuracy is reached. In his remarkable
thesis, Orin Merrill describes his independent discovery of a
similar technique. Merrill takes the opposite ends of the
cylinder described above to be close together, so that each
simplex in the decomposition of the cylinder touches both
ends. His method (given the illuminating name of the "sandwich" method by later writers) therefore moves from the initial guess to the opposite end of the cylinder very rapidly
and as a consequence must be content, in each iteration, with
a modest refinement of the previous guess. This approximation, however, may be taken as an initial guess for a subsequent round of the algorithm; the basic cycle is then repeated until a satisfactory accuracy is obtained.

Several variations of these two methods have been programmed and the substantial computational experience obtained
by Merrill, Kuhn, Wilmuth and others indicate that they perform remarkably well in contrast to the older techniques
which had previously been available. I have not yet seen an
explicit comparison to the naive approach in which a rough
guess obtained by a fixed point technique is refined by
Newton's method, but I feel sure the algorithms of Eaves and
Merrill will remain one of the major approaches to fixed
point computations in the years to come.

During the last decade all of the fixed point methods
with which I had been familiar were based upon a decomposition of the simplex into combinatorial objects such as subsimplices or primitive sets. These objects permit a replacement operation which allows a discrete movement from an initial guess to a final approximation. The difficulties involved in solving a system of non-linear equations and inequalities by directly tracing out a path leading to the
solution were replaced by a combinatorial approach to
Brouwer's theorem with a constructive flavor.

In the 1950's, however, it had become apparent to mathematicians that many of the intricate arguments of combinatorial topology could be replaced by constructions which were
simpler and more intuitive, if adequate differentiability assumptions were placed on the underlying manifolds and functions specifying the particular problem. For example, in a
paper written in 1963 Morris Hirsch gives an elegant proof of
Brouwer's theorem involving simplicial subdivisions, which
has many points of similarity to our computational procedures.
In a final paragraph he remarks that an alternative proof can

be based on the concept of "regularity" of a differentiable
map.

For many of us one of the great surprises of the conference at Clemson was the paper by Kellogg, Li and Yorke
which presented the first computational method for finding a
fixed point of a continuous mapping making use of the considerations of differential topology instead of our customary
combinatorial techniques. In this paper the authors show how
Hirsch's argument can be used to define paths leading from
virtually any pre-assigned boundary point of the simplex to a
fixed point of the mapping. Stephen Smale has also communi-
cated to me recently the results of a similar study analyzing, in detail, the systems of differential equations which
arise in this fashion. Both Smale and the three authors mentioned above make the important observation that the path
which is being calculated by their methods -- near the fixed
point -- is virtually identical with that which would be followed were Newton's method being used.

In order to explain why I find this observation important, let me begin by saying that the differentiable methods
consist essentially of tracing the solutions of a system of
non-linear equations, say $F(x) = c$, which typically involve one less equation than unknown. Under suitable assumptions this set of solutions forms a one-dimensional differentiable manifold which, by the proper selection of the
constant c , will have a component leading from the pre-
assigned boundary point to a desired fixed point. But as
Eaves and I have shown in a recent paper, virtually all of
the simplicial methods which have dominated the horizon during the last decade can be put in an identical form with the
function F being <u>piecewise</u> <u>linear</u> rather than

differentiable. In other words, there is essentially no distinction between the differentiable methods and those we have
been in the habit of using, at least in their general out-
lines. (There are, of course, technical distinctions: differential equations may have great advantages over difference
equations, since the step size for the former need not be assigned in advance -- on the other hand a discrete system may
more easily accommodate mappings which are not smooth).

The great similarity between the combinatorial and differentiable algorithms suggests to me the possibility that a
discrete method -- such as Eaves' -- in which the grid size
is continually decreasing, may very well be behaving like
Newton's method near the solution. If this argument could be
made precise, it might provide an explanation for the unex-
pectedly small number of iterations typically required by
fixed point methods, and their virtually linear behavior in
the neighborhood of the solution. It might also suggest a
similarity between the sophisticated methods based on homotopy arguments and the naive approaches in which a fixed
point approximation is followed by a conventional Newton's
method. The paper by Fisher, Gould and Tolle, presented
here, may represent a step in the direction of understanding
these phenomena.

A number of authors, including Shapley, Lemke, Kuhn,
and Scarf and Eaves have recently been concerned with the application of the topological concepts of index theory to
fixed point methods. It is possible to associate with each
solution of a particular problem -- say a completely labelled
simplex -- an index which is either +1 or -1 . The index
is defined fully in terms of the data specifying the solution
and is independent of the path which has been used in

arriving at that solution. For example, consider a problem whose underlying combinatorial structure is given by a simplicial subdivision of the simplex

$$\left\{ x = (x_1,\ldots,x_n) \mid x_i \geq 0 \quad, \quad \sum_1^n x_i = 1 \right\}$$

In the special case of integer labelling, each vertex of the subdivision will receive an integer label from 1 to n , with a special provision for those vertices on the boundary. The index of a completely labelled simplex $(v^1,\ldots,v^n)$ is then defined as follows: order the vertices by requiring v^1 to have the label $1, v^2$ the label 2 , etc. The index of the solution is then +1 if the determinant of $[v_j^i]$ is positive and −1 if this determinant is negative. A corresponding definition can also be provided for vector labels -- one simply conpares the sign of the above determinant with that of the determinant whose columns are composed of the associated vector labels. If the two agree in sign the index is +1 ; otherwise it is −1 .

In the context of fixed point methods the major conclusion of index theory is that the number of solutions to a given problem with positive index exceeds the number of solutions with negative index by one. Of course, this result implies, but is much stronger than the simple statement that the number of solutions is odd.

For example, uniqueness of the solution could be demonstrated if it were possible to show that all solutions to a particular problem necessarily had a positive index. This line of argument, available in both discrete and differentiable forms, is capable of demonstrating the major theorems concerning the uniqueness of the competitive equilibrium in a general Walrasian model of exchange. I believe that it may also provide an alternative proof of the Gale-Nikaido uniqueness theorems. It may indeed, be correct that virtually every theorem in mathematical economics and operations research asserting uniqueness will turn out to be essentially a statement that the solutions to a given problem all have a positive index.

Even though my above remarks have stressed the recent theoretical developments of greatest interest to me, I should like to suggest that the ultimate test of our collective efforts in this field must reside in the ability to solve specific problems which have been resistant to earlier techniques. Many of the fixed point methods which have been developed over the last ten years have been tested on artificial problems constructed primarily to explore the characteristics of their performance. But we have also seen, in the work of Allgower and his colleagues on differential equations, and in the papers by Shoven, Whalley, and others on the construction of models of general equilibrium, that our methods can be applied to realistic and important problems. It is my hope that increasing numbers of applications will be seen in the future, drawing on the many theoretical contributions which will continue to be made by all of us working in this field of research.

Finding Roots of Polynomials By Pivoting

Harold W. Kuhn

ABSTRACT

A constructive algorithm is given that generates sequences that converge to all of the roots of a polynomial with complex coefficients. It is based on a pivoting procedure that operates on labelled triangulations of the complex plane.

1. Introduction and Summary

A new proof of the Fundamental Theorem of Algebra based on a labelling rule and a combinatorial lemma applying to labelled triangulations of the complex plane was presented in [1]. Although the purpose of that paper was limited strictly to the proof of existence, the constructive nature of the argument suggested algorithmic computation of the roots by the same procedure. Both the existence proof and the algorithm to be given in this paper were influenced by the recent success of combinatorial algorithms of Lemke and Howson ([2] and [3]) for the linear complementarity problem and of Scarf [4] for the approximation of Brouwer fixed points. The work reported here is, in many ways, an ideal example of the subsequent development of these techniques. It uses in an essential way two ideas that are central to the current implementation of pivotal methods for the approximation of solutions of nonlinear systems. These are:

(1) Artificial labelling to initiate the algorithm using a simple problem homotopically equivalent to the given problem (see [5], [6], and [7]).

(2) Repeated homotopic subdivision using piecewise linear interpolations (see [8] and [9]).

In Section 2, a labelling rule for general triangulations of the complex plane is defined and several fundamental results linking the labelling with the roots of the polynomial are established. The special simplicial subdivision (of half of 3-space into tetrahedra) used in the algorithm is

described in Section 3.   The algorithm is given a geometric
statement in Section 4.   Precisely, the labelling of Section
2 applied to the special triangulations of Section 3 defines
a directed graph with as many unbounded components as the
polynomial has roots.   These components in turn define sequences in the complex plane.   Section 5 contains the proof
that these sequences converge to the roots of the polynomial.

2.   Labelled Triangulations

Let f(z) be a monic polynomial of degree n in the
complex variable z with complex numbers as coefficients,
that is, $f(z) = z^n + a_1 z^{n-1} + \ldots + a_n$, where $a_1, \ldots, a_n$ are
complex constants. We seek the n roots of f(z), counting
multiplicities; if these are denoted by $\bar{z}_1, \ldots, \bar{z}_n$ in some
order, then $f(z) = (z - \bar{z}_1) \cdots (z - \bar{z}_n)$. The polynomial f(z)
induces a <u>labelling</u> of the points of the z-plane using the
labels 1, 2, and 3.

<u>Definition 2.1</u>:   If $f(z) = u + iv \neq 0$, choose
arg f(z) to be the unique angle α such that $-\pi < \alpha \leq \pi$
and $\cos \alpha = u/(u^2 + v^2)^{1/2}$, $\sin \alpha = v/(u^2 + v^2)^{1/2}$.   Then
$\ell(z)$, <u>the</u> <u>label</u> <u>of</u> z <u>induced</u> <u>by</u> f , is defined by:

$$
\ell(z) = \begin{cases} 1 & \text{if} & -\pi/3 \leq \arg f(z) \leq \pi/3 & \text{or} & f(z)=0 \\ 2 & \text{if} & \pi/3 < \arg f(z) \leq \pi \\ 3 & \text{if} & \pi < \arg f(z) < -\pi/3 \ . \end{cases}
$$

Definition 2.1 is illustrated geometrically in Figure
2.1.   The label of z is determined by the sector in which
f(z) falls.   Open boundaries are indicated by dotted lines.

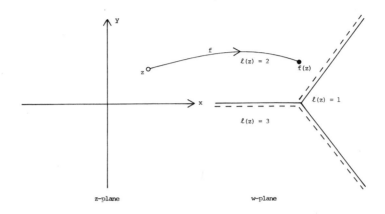

Figure 2.1

As in [1], we shall seek very small triangles in the
plane that carry all three labels. In subsequent sections,
these triangles may degenerate to three collinear points.
Therefore, we shall speak of completely labelled triples
rather than triangles. Moreover, on some occasions, the labels will be induced by either f(z) or another polymonial
g(z) , with some points labelled by one polynomial and the
remaining labelled by the other.

Definition 2.2: The triple $\{z_1,z_2,z_3\}$ is completely
labelled by f or g if $\ell(z_k) = k$ and is induced by
$f(z_k)$ or $g(z_k)$ for k = 1,2,3 . If all of the labels are
induced by f , we shall say that the triple is completely
labelled by f .

Note that, when a triple is completely labelled by f
or g , it is possible that two of the points are the same.
The following results establish connections between

completely labelled triples and the roots of $f(z)$. The first was proved in [1], p. 149, and so the demonstration will not be reproduced here.

Proposition 2.1: If $\{z_1, z_2, z_3\}$ is completely label- led by f and $|f(z_j) - f(z_k)| \leq \varepsilon$ for $1 \leq j < k \leq 3$ then $|f(z_k)| \leq 2\varepsilon/\sqrt{3}$ for $k = 1, 2, 3$.

Definition 2.2: For any triple $\{z_1, z_2, z_3\}$,

$$\text{diam } \{z_1, z_2, z_3\} = \max_{1 \leq j < k \leq 3} |z_j - z_k| \; .$$ If T is any triangula- tion of the plane, the mesh of T is denoted by $\mu(T)$ and equals sup diam $\{z_1, z_2, z_3\}$ over all triangles in T .

Proposition 2.2: Suppose T is a triangulation of the plane with $\mu(T) = \mu < \infty$. Then there exists R such that there is no triangle of T completely labelled by f out- side $|z| = R$. In general, R will depend on f and μ . However, for small enough μ , any $R > \max_k |a_k| + 1$ will do.

Proof. Rewrite $f(z) = z^n (1 + g(z))$. Then

$$|g(z)| \leq \frac{|a_1|}{R} + \ldots + \frac{|a_n|}{R^n} \leq \frac{\max_k |a_k|}{R-1} \text{ for } |z| \geq R > 1 \; .$$

For $R > \max_k |a_k| + 1$, set $\lambda = 1 - \dfrac{\max_k |a_k|}{R-1}$. Then

$0 < \lambda \leq 1$ and $|f(z)| \geq \lambda |z|^n$ for $|z| \geq R$. On the other hand, let $\{z_1, z_2, z_3\}$ be a triangle of T completely label- led by f with $|z_\ell| = \max_k |z_k| \geq R$. Then

$$|f(z_j)-f(z_k)| \leq \mu\{n|z_\ell|^{n-1}+(n-1)|a_1||z_\ell|^{n-2}+\cdots+|a_{n-1}|\}$$

$$\leq \mu n^2 M|z_\ell|^{n-1}$$

for $M = \max\{1,|a_1|,\cdots,|a_{n-1}|\}$ and $1 \leq j < k \leq 3$. By Proposition 2.1, this implies

$$|f(z_\ell)| \leq \frac{2}{\sqrt{3}} \mu n^2 M|z_\ell|^{n-1} \quad .$$

Combining these two estimates, we have

$$|z_\ell| \leq \frac{2\mu n^2 M}{\sqrt{3}\,\lambda}$$

for $R > \max_k |a_k| + 1$. Note that n^2 and M depend only on $f(z)$ and that λ increases with R . We also know $|z_\ell| \geq R$ by assumption. Hence R can be chosen large enough (depending on μ) to obtain a contradiction. On the other hand, for any fixed $R > \max_k |a_k| + 1$, a contradiction is obtained when μ is small enough. ∎

   The result says that, for triangulations T of mesh $\mu(T) \leq \mu_o$, there are no triangles of T completely labelled by f outside a bounded region of the plane, which depends only on μ_o . The next proposition asserts that the only triangles completely labelled by f occur "near" roots of f . (Note that it is stated in terms of triples, which may be collinear.)

Proposition 2.3:   If $\{z_1, z_2, z_3\}$ is a triple with diam $\{z_1, z_2, z_3\} = \delta$ completely labelled by f then some root of f is closer to the triple than $6n\delta/\pi$.

Proof: Suppose $f(z_k) = w_k$ for $k = 1,2,3$.   Then, by the labelling rule:

$$0 < \arg w_1 - \arg w_3 < 4\pi/3$$
$$0 < \arg w_2 - \arg w_1 < 4\pi/3$$
$$2\pi/3 < \arg w_2 - \arg w_3 \quad .$$

Hence, either

$$\pi/3 < \arg w_1 - \arg w_3 < 4\pi/3$$

or

$$\pi/3 < \arg w_2 - \arg w_1 < 4\pi/3 \quad .$$

Restating this, if there is a completely labelled triple of diameter δ than there exist z and z' with $|z'-z| \leq \delta$ and $\pi/3 < \arg f(z') - \arg f(z) < 4\pi/3$. If no root of f is closer to the triple than $6n\delta/\pi$ then $|z-\bar{z}_j| \geq 6n\delta/\pi$ for $j = 1,\ldots,n$. Hence, $|(z'-z)/(z-\bar{z}_j)| \leq \pi/6n \leq 1$ for all j . It was shown in [1], p. 155, that $|w| \leq 1$ implies $|\arg (1+w)| < 2|w|$. Therefore $|\arg (z'-\bar{z}_j)/(z-\bar{z}_j)| < \pi/3n$ for all j , which implies $|\arg f(z')/f(z)| < \pi/3$ and either

$$\arg f(z') - \arg f(z) < \pi/3$$

or

$$\arg f(z') - \arg f(z) > 5\pi/3 \quad .$$

This contradicts the bounds established above for this difference and proves the proposition. ▮

Of course, Proposition 2.3 has the simple corollary, already established as a consequence of Proposition 2.2, that there are no triangles completely labelled by f of a triangulation T of fixed mesh outside a bounded region of the plane. The following result establishes this same conclusion when the labelling is done by either of two polynomials of degree n, f or g .

Proposition 2.4: Let $f(z)$ and $g(z)$ be two monic polynomials of degree n . If $\{z_1, z_2, z_3\}$ is a triple with diam $\{z_1, z_2, z_3\} = \delta$ completely labelled by f or g then there exists $r > 0$ (depending on δ and the roots of f and g) such that some root of f is closer to the triple then r .

Proof: The proof follows that of Proposition 2.3 except that we must handle the case of z' labelled by g and z labelled by f , where $\pi/3 < \arg g(z') - \arg f(z) < 4\pi/3$. In this case, we write

$$\frac{z'-\bar{z}_j}{z-\bar{z}_j} = 1 + \frac{z'-z}{z-\bar{z}_j} + \frac{\bar{z}_j-\bar{z}_j'}{z-\bar{z}_j}$$

By choosing r large enough, we can insure

$$\left| \frac{z'-z}{z-\bar{z}_j} + \frac{\bar{z}_j-\bar{z}'_j}{z-\bar{z}_j} \right| \leq \frac{\delta+\max_k |\bar{z}_k-\bar{z}'_k|}{|z-\bar{z}_j|} \leq \frac{\pi}{6n}$$

if $|z-\bar{z}_j| \geq r$ for all j . Hence $|\arg g(z')/f(z)| < \pi/3$ and this leads to the same contradiction as before. ∎

3. A Special Triangulation

The half of a 3-space expressed as

$$C \times [-1,\infty) = \{(z,d)\,|\,z=x+iy \in C , \quad d \geq -1\}$$

contains a sequence of replicas C_d of the complex plane C for $d = 1,0,1,2,\ldots$. Given a <u>center</u> $\bar{z}$ and a <u>grid size</u> $h > 0$, we define a triangulation $T_d(\bar{z};h)$ of each of these planes. The triangulation $T_{-1}(\bar{z};h)$ is illustrated in Figure 3.1.

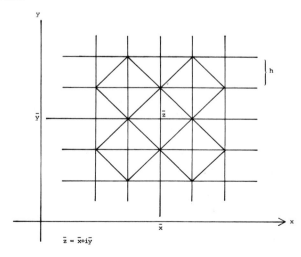

Figure 3.1

A triangle in $T_{-1}(\bar{z};h)$ is specified uniquely by a pair of

integers (r,s) with $r + s$ <u>even</u> and

$(a,b) \in \{(1,0),(0,1),(-1,0),(0,-1)\}$. The z-coordinates of

its vertices are:

$$[\bar{x} + rh] \quad + \quad i[\bar{y} + sh]$$
$$[\bar{x} + (r+a)h] \quad + \quad i[\bar{y} + (s+b)h]$$
$$[\bar{x} + (r-b)h] \quad + \quad i[\bar{y} + (s+a)h] \quad .$$

The mesh of the triangulation is $\sqrt{2} \, h$.

The triangulation $T_d(\bar{z};h)$ is illustrated in Figure

3.2, where $d = 0,1,2,\ldots$.

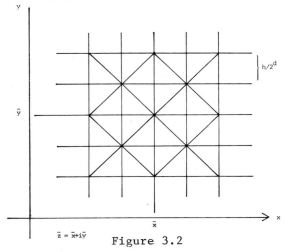

$$\bar{z} = \bar{x} + i\bar{y}$$

Figure 3.2

A triangle in $T_d(\bar{z};h)$, where $d \geq 0$, is specified uniquely

by a pair of integers (r,s) with $r + s$ <u>odd</u> and

$(a,b) \in \{(1,0),(0,1),(-1,0),(0,-1)\}$. The z-coordinates of

its vertices are:

$$[\bar{x} + rh/2^d] \qquad + \quad i[\bar{y} + sh/2^d]$$

$$[\bar{x} + (r+a)h/2^d] \quad + \quad i[\bar{y} + (s+b)h/2^d]$$

$$[\bar{x} + (r-b)h/2^d] \quad + \quad i(\bar{y} + (s+a)h/2^d] \quad .$$

The mesh of the triangulation is $\sqrt{2}\, h/2^d$.

The layer between C_{-1} and C_o is subdivided into
tetrahedra as shown in Figure 3.3, which shows the subdivision above a square centered at $(\bar{x} + rh) + i(\bar{y} + sh)$ with
<u>both</u> r and s <u>even</u>. All similar squares are treated in
like manner.

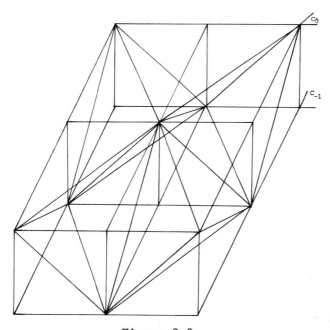

Figure 3.3

The layer between C_d and C_{d+1} , where $d \geq 0$, is
subdivided into tetrahedra as shown in Figure 3.4, which
shows the subdivision above a square in C_d with vertices at

$$[\bar{x} + rh/2^d] \qquad + i[\bar{y} + sh/2^d]$$

$$[\bar{x} + (r+a)h/2^d] \quad + i[\bar{y} + sh/2^d]$$

$$[\bar{x} + (r+a)h/2^d] \quad + i[\bar{y} + (s+b)h/2^d]$$

$$[\bar{x} + rh/2^d] \qquad + i[\bar{y} + (s+b)h/2^d]$$

where <u>both</u> r and s are <u>even</u> and
(a,b) **e** $\{(1,1),(-1,1),(-1,-1),(1,-1)\}$. All similar squares
are treated in like manner.

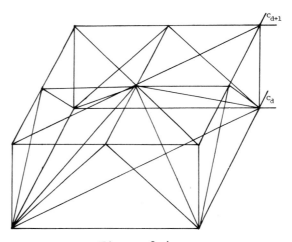

Figure 3.4

Note that no new vertices have been added to define this
subdivision.

In the algorithm to be presented in the next section,
we shall be concerned with triples of points
$\{(z_1,d_1),(z_2,d_2),(z_3,d_3)\}$ which are the vertices of a triangular face of a tetrahedron of this subdivision. Note that

d_k = d or d + 1 for some fixed d $\geq$ -1 and k = 1,2,3 .
In this event, we shall say that the triple $\{z_1,z_2,z_3\}$ lies between levels d and d + 1 .

   Proposition 3.1: If the triple $\{z_1,z_2,z_3\}$ lies between levels d and d + 1 then

$$\text{diam } \{z_1,z_2,z_3\} \leq \sqrt{2} \ h/2^d .$$

   Proof: The diameter of all possible triples lying between levels d and d + 1 is easily computed from Figures 3.3 and 3.4.   ∎

4. The Algorithm

   Each of the vertices in the subdivision defined in Section 3 is assigned a label 1, 2, or 3. For vertices of $T_{-1}(\bar{z};h)$, this labelling is induced by the polynomial $(z-\bar{z})^n$. For vertices of $T_d(\bar{z};h)$, where d = 0,1,... , this labelling is induced by the polynomial f(z) . The labelling allows us to define a directed graph G associated with the labelled subdivision. This will be done in three steps, corresponding to dimensions 1, 2, and 3.
   In the plane C_{-1} triangulated by $T_{-1}(\bar{z};h)$, let $Q_m(\bar{z};h)$ denote the square with corners at $(\bar{x} \pm mh)$ + $i(\bar{y} \pm mh)$, where m is a positive integer. The boundary $\partial Q_m(\bar{z};h)$ is oriented counterclockwise and the triangles of $T_{-1}(\bar{z};h)$ inside $Q_m(\bar{z};h)$ are oriented in the customary counterclockwise cyclic order of their vertices.

Definition 4.1: A node of the digraph G_1 is an edge
on $\partial Q_m(\bar{z};h)$ with at least one label of 1 on its vertices.
A directed edge joins two nodes if the corresponding edges
meet in a vertex labelled 1 . The edge is directed to
coincide with orientation of $\partial Q_m(\bar{z};h)$.

Figure 4.1 shows G_1 for n = 3 and m = 2 .

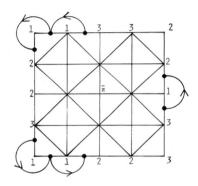

Figure 4.1

It is evident that the components of G_1 start at
edges of $\partial Q_m(\bar{z};h)$ labelled (2,1) or (3,1) and end at
edges labelled (1,2) or (1,3) .

Proposition 4.1: If $m \geq 3n/2\pi$ then G_1 has n components starting at edges labelled (3,1) and ending at
edges labelled (1,2) on $\partial Q_m(\bar{z};h)$. For such m , there
are no edges labelled (2,1) on $\partial Q_m(\bar{z};h)$.

Proof: As shown in [1], p. 154,

$$\frac{n}{2m} < \arg \frac{(z_2-\bar{z})^n}{(z_1-\bar{z})^n} < \frac{n}{m}$$

for edges (z_1, z_2) e $\partial Q_m(\bar{z}; h)$ and $m \geq n/\pi$. By the labelling rule, if $n/m \leq 2\pi/3$ then no edge is labelled $(1,3)$,
$(3,2)$, or $(2,1)$. Consider the n disjoint open sectors
of angle $2\pi/3n$ in C_{-1} centered at $\bar{z}$ of points labelled
3. By the choice of m , some vertex of $\partial Q_m(\bar{z}; h)$ falls in
each such sector and is given the label 3. The first succeeding vertex on $\partial Q_m(\bar{z}; h)$ outside the sector must be given
the label 1 and the Proposition follows.   ∎

Definition 4.2: A <u>node</u> of the digraph G_2 is a triangle of $T_{-1}(\bar{z}; h)$ inside $Q_m(\bar{z}; h)$ with labels 1 and 2
on at least one of its sides. A <u>directed</u> <u>edge</u> joins two
nodes if the corresponding triangles meet in a side with labels 1 and 2. The edge is directed so that the label 1 is on
the left and the label 2 is on the right.

It is evident that the components of G_2 <u>start</u> either
at triangles with a side labelled $(1,2)$ on $\partial Q_m(\bar{z}; h)$ or at
triangles labelled $(1,3,2)$ in counterclockwise order and
<u>end</u> either at triangles with a side labelled $(2,1)$ on
$\partial Q_m(\bar{z}; h)$ or at triangles labelled $(1,2,3)$ in counterclockwise order. Figure 4.2 shows G_2 for $n = 3$ and $m = 2$.

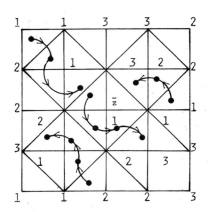

Figure 4.2

Proposition 4.2: If $m \geqq 3n/2\pi$ then G_2 has n components starting at triangles labelled (1,2) on $\partial Q_m(\bar{z};h)$ and ending at triangles labelled (1,2,3) inside $Q_m(\bar{z};h)$. The other components of G_2, if any, connect triangles labelled (1,3,2) to triangles labelled (1,2,3) inside $Q_m(\bar{z};h)$.

Proof: This is an immediate corollary to Proposition 4.1 combined with the observations made above. ∎

Definition 4.3: A node of the digraph G_3 is a tetrahedron of the subdivision of $C \times [-1,\infty)$ with labels 1, 2, and 3 on at least one of its faces. A directed edge joins two nodes if the corresponding tetrahedra have a common face with all three labels. The edge is directed as a right-handed screw rotated in the cyclic order of the labels (1,2,3). (This is illustrated in Figure 4.3.)

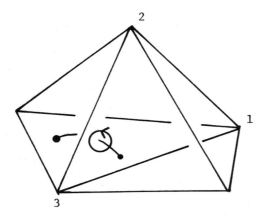

Figure 4.3

(For readers who may find this definition too geometri-
cal, let $V_k = (x_k + iy_k, d_k)$ be the common vertices of the two
tetrahedra with the label of $V_k = k$ for $k = 1,2,3$. If
the two additional vertices are $V_4 = (x_4 + iy_4, d_4)$ and
$V_4' = (x_4' + iy_4', d_4')$ then the signs of the determinants

$$
\begin{vmatrix}
x_1 & y_1 & d_1 & 1 \\
x_2 & y_2 & d_2 & 1 \\
x_3 & y_3 & d_3 & 1 \\
x_4 & y_4 & d_4 & 1
\end{vmatrix}
\quad \text{and} \quad
\begin{vmatrix}
x_1 & y_1 & d_1 & 1 \\
x_2 & y_2 & d_2 & 1 \\
x_3 & y_3 & d_3 & 1 \\
x_4' & y_4' & d_4' & 1
\end{vmatrix}
$$

are opposite since V_4 and V_4' lie on opposite sides of the
plane through V_1 , V_2 , and V_3 . The directed edge goes
from the negative tetrahedron to the positive tetrahedron.)

From the definition given above, a node of G_3 may
have one edge in, one edge out, or both one edge in and one
edge out. The nodes with only one edge in or one edge out

must correspond to tetrahedra between C_{-1} and C_o that have a face with all three labels in C_{-1} , that is, to starts or ends of components of G_2 if we can be certain that all completely labelled triangles in $T_{-1}(\bar{z};h)$ lie inside $Q_m(\bar{z};h)$. If this underlined condition is satisfied, and the conclusion of Proposition 4.1 is valid, then the appropriate components of G_1 , G_2 , and G_3 can be connected (ends to starts) to construct n unbounded components starting on $\partial Q_m(\bar{z};h)$.

Proposition 4.3: If $m \geq 3(1+\sqrt{2})n/4\pi$ then there is no triangle of $T_{-1}(\bar{z};h)$ completely labelled by $(z-\bar{z})^n$ outside $Q_m(\bar{z};h)$.

Proof: The choice of m in this proposition is made so that, if (z,z') is an edge of the triangulation lying outside or on the boundary of $Q_m(\bar{z};h)$ then it can be oriented so that

$$0 < \arg (z'-\bar{z})/(z-\bar{z}) < 2\pi/3n \quad .$$

The argument of Proposition 4.1 establishes this fact for horizontal and vertical edges since $3(1+\sqrt{2})n/4\pi > 3n/2\pi$. Hence it must be proved only for the diagonals of the triangulation. The worst case is shown in Figure 4.4, where k is to be chosen so as to maximize α .

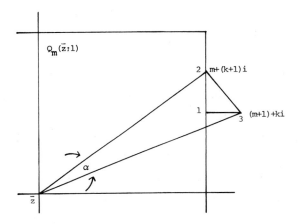

Figure 4.4

Elementary calculations show that

$$\tan \alpha = \frac{m + k + 1}{m^2 + m + k^2 + k}$$

and this is maximized (over continuous values of k) when

$k = \sqrt{2m(m+1)}\ -m-1$ with the value $\tan \alpha = 1/(2k+1)$. Since the maximum value over integral k cannot be greater, we have

$$\alpha < \tan \alpha \underset{=}{<} \frac{1}{2\sqrt{2m(m+1)}\ -2m-1} < \frac{1+\sqrt{2}}{2m}$$

provided $\alpha < \pi/2$. Finally, the choice of $m \geq 3(1+\sqrt{2})n/4\pi$ guarantees $0 < \alpha < 2\pi/3n$ as desired.

To complete the proof of the proposition, suppose we have a completely labelled triangle outside $Q_m(\bar{z};h)$. We may assume that the vertices are   z, z´, and z´´   with

$$0 < \arg (z´-\bar{z})/(z-\bar{z})   < 2\pi/3n$$
$$0 < \arg (z´´-\bar{z})/(z´-\bar{z}) < 2\pi/3n$$
$$0 < \arg (z´´-\bar{z})/(z-\bar{z})   < 2\pi/3n   .$$

If $\ell(z) = \ell$, then the first inequality implies $\ell(z´) = \ell+1$ (the label following ℓ the cyclic order 1,2,3) .
Since the triangle is completely labelled, $\ell(z´´) = \ell+2$
(the label preceding ℓ in the cyclic order 1,2,3) and
this contradicts the third inequality.   ▌

Definition 4.4: Let $m \geq 3(1+\sqrt{2})n/4\pi$, $\bar{z}$, and h
be given. The digraph G has   n   components, $(V_{j1}, V_{j2}, \ldots)$
for $j = 1, \ldots, n$, which will be defined recursively. All
nodes belong to G_1 , G_2 , or G_3 . The initial nodes V_{j1}
are the   n   edges of $\partial Q_m(\bar{z};h)$ labelled (3,1) . (There
are exactly   n   such by Proposition 4.1 and the choice of
m.) If the node $V_{jk} \in G_i$ then $V_{j,k+1}$ is the following
node in G_i , if there is one. Otherwise, the following
rules are followed:

(a) If $V_{jk} \in G_1$ and labelled (1,2) then $V_{j,k+1}$
is the unique node in G_2 containing V_{jk} .

(b) If $V_{jk} \in G_2$ and is labelled (1,2,3) then
$V_{j,k+1}$ is the unique node in G_3 containing V_{jk} .

(c) If $V_{jk} \in G_3$ and has a face labelled (1,3,2) in
$T_{-1}(\bar{z};h)$ then $V_{j,k+1}$ is this node in G_2 . (This triangle is inside $Q_m(\bar{z};h)$ by Proposition 4.3.)

For the purposes of the convergence proof to be given
in the next section, it is more convenient to deal with sequences of values of z than with components of G . However, this translation is easily made in view of the following observations:

In each component of G_1, G_2, or G_3 successive nodes

exhibit the combinatorial pattern of "pivoting", namely, one
vertex is dropped and a new vertex enters. It is this new
vertex that continues the sequence. When two components are
connected together, increasing dimension (that is, from G_1

to G_2 or from G_2 to G_3) the second node has one new ver-

tex that is not in the first node. This new vertex continues
the sequence. When two components are connected decreasing
dimension (G_3 to G_2), a vertex is dropped when passing from

the first node to the second node. However, the next node in
G is again in G_2 and, by a previous remark, has a new ver-

tex which is to be added to the sequence.

Definition 4.5: Let $m \geq 3(1+\sqrt{2})n/4\pi$, $\bar{z}$, and h de-

fine the n components (V_{jk}) of G . These define n

sequences (z_{jk}, d_{jk}) recursively. Let (z_{j1}, d_{j1}) be the

endpoint of V_{j1} . Suppose $(V_{j1}, V_{j2}, \ldots, V_{jk'})$ has been

used to define $((z_{j1}, d_{j1}), \ldots, (z_{jk}, d_{jk}))$ where $k \leq k'$.

Then $(z_{j,k+1}, d_{j,k+1})$ is the unique vertex in $V_{j,k'+1}$ but

not in $V_{jk'}$ unless $V_{jk'} \in G_3$ and $V_{j,k'+1} \in G_2$. In the

latter case $V_{j,k'+2} \in G_2$ and $(z_{j,k+1}, d_{j,k+1})$ is the

unique vertex in $V_{j,k'+2}$ but not in $V_{j,k'+1}$. (Note

that, by Proposition 4.3, this exceptional case can only

occur a finite number of times and so the recursion produces
n infinite sequences $((z_{jk}, d_{jk}))$ where $j = 1,\ldots,n$ and
$k = 1, 2, \ldots$.)

It is important to note that, although the nodes of G
are all distinct, the same (z,d) may appear several times
in the same sequence or may appear in different sequences.
We shall list the properties of these sequences that follow
from the properties of G and which will be used in the convergence proof.

Proposition 4.4: Given $d \geq 0$, there exists an integer K such that, for $k \geq K$, each (z_{jk}, d_{jk}) belongs to
a triple $\{(z_{jk}, d_{jk})$, $(z_{jk'}, d_{jk'})$, $(z_{jk''}, d_{jk''})\}$ completely labelled by f , where $k > k', k''$ and
$d' \leq d_{jk}$, $d_{jk'}$, $d_{jk''} \leq d' + 1$ for some $d' \geq d$.

Proof: The nodes in G that correspond to edges, triangles, and tetrahedra in $C \times [-1,d]$ are finite in number
by Propositions 2.3 and 2.4. Hence, after finite initial
segments of the n sequences, all (z_{jk}, d_{jk}) are vertices
of nodes V of G that correspond to tetrahedra in
$C \times [d, \infty]$.   ∎

Proposition 4.5: For fixed m , $\bar{z}$, and h , the terms
$z_{j,k+1}, d_{j,k+1})$ are determined uniquely by
$((z_{j1}, d_{j1}), \ldots, (z_{jk}, d_{jk}))$ and the labels given these points
by f .

Proof: This follows directly from Definitions 4.1-
4.5.   ∎

5. Convergence Proof

With a choice of $m \geq 3(1+\sqrt{2})n/4\pi$, $\bar{z}$, and h , the
algorithm described in Section 4 applied to $f(z)$ produces
n sequences of values of z which have been denoted by
(z_{jk}) where $j = 1,\ldots,n$ and $k = 1,2,\ldots$. (The sequences are uniquely indexed by j if a fixed order is
chosen for the n initial nodes of G on $\partial Q_m(\bar{z};h)$; note
that this choice is independent of f .)

Theorem 5.1: The n sequences (z_{jk}) converge to the
n roots $\bar{z}_1,\ldots,\bar{z}_n$ of $f(z)$. These roots can be ordered
so that

$$\lim_{k \to \infty} z_{jk} = \bar{z}_j \qquad \text{for } j = 1,\ldots,n .$$

Proof: Given the preparation of the previous sections,
we are very close to this result. By Proposition 4.4, for
large enough k , z_{jk} belongs to a completely labelled triple above a preassigned level d . By Proposition 3.1, this
triple is small, and by Proposition 2.3, z_{jk} lies close to
some root. Since the locations of roots are separated, every
sequence z_{jk} converges to some root of $f(z)$. However,
this does not prove that all of the roots are the limits of
some sequence (and with the correct multiplicities in case
there are multiple roots). The demonstration of this fact
will depend on a reduction of the general case to the special
case of polynomials with all roots distinct.

Note that $f(z) + \varepsilon$ has the derivative $f'(z)$ for all $\varepsilon > 0$. Let the roots of $f'(z)$ be $z'_1, \ldots, z'_{n-1}$. Then $f(z) + \varepsilon$ has all roots distinct if $\varepsilon \neq -f(z'_j)$ for $j = 1, \ldots, n-1$. Hence, avoiding this finite set of values, we can choose ε_o such that $f(z) + \varepsilon$ has all roots distinct for $0 < \varepsilon \leqq \varepsilon_o$. Let the n sequences produced by the algorithm using $f(z) + \varepsilon$ for labelling be denoted by $(z_{jk}(\varepsilon))$ where $j = 1, \ldots, n$ and $k = 1, 2, \ldots$. (The index j is assigned using the previous ordering of the n starts on $\partial Q_m(\bar{z}; h)$ which is independent of f and ε.) Remarkably, the sequences $(z_{jk}(\varepsilon))$ can be made to coincide with (z_{jk}) as long as we please; this is expressed in the following lemma.

Lemma 5.1: Let $f(z)$, $m \geq 3(1+\sqrt{2})n/4\pi$, $\bar{z}$, h be given and define (z_{jk}) and $(z_{jk}(\varepsilon))$ as in Definition 4.5. Given K, there exists ε_1 such that $z_{jk}(\varepsilon) = z_{jk}$ for $1 \leq k \leq K$ and $0 < \varepsilon \leqq \varepsilon_1$.

Proof: As noted in Proposition 4.5, the sequence (z_{jk}) is determined uniquely by the labels assigned by $f(z)$. For the finite set of values $(z_{j1}, \ldots, z_{jK})$, where $j = 1, \ldots, n$, the labels given by $f(z)$ and $f(z) + \varepsilon$ will be the same for $\varepsilon > 0$ small enough since the regions of the w-plane in Figure 2.1 yielding labels 1, 2, and 3 are open to the right. ∎

Corollary 5.1:

$$\lim_{k \to \infty} \lim_{\varepsilon \to 0+} z_{jk}(\varepsilon) = \lim_{k \to \infty} z_{jk}.$$

We shall now treat the behavior of the algorithm for polynomials with all roots distinct. For such a polynomial, let $\rho(f) = \frac{1}{2} \max_{j \neq \ell} |\bar{z}_j - \bar{z}_\ell|$. Then the open neighborhoods $N_\rho(\bar{z}_j) = \{z \mid |z - \bar{z}_j| < \rho\}$ are disjoint for $0 < \rho < \rho(f)$. Finally, for the special purpose of the following lemma, let $\alpha(T)$ be the supremum of all angles in triangles of a triangulation T .

Lemma 5.2: Let $f(z)$ be a polynomial with all roots distinct. Given ρ , $0 < \rho < \rho(f)$, there exists exactly one triangle inside each $N_\rho(\bar{z}_j)$ completely labelled by f for each triangulation T with $\mu(T)$ small enough and $\alpha(T) < 2\pi/3$.

Proof: By Proposition 2.3, all triangles completely labelled by f lie inside some $N_\rho(\bar{z}_j)$ and arbitrarily close to $\bar{z}_j$ for $\mu(T)$ small enough. Since all of the roots of f are simple, the mapping defined by f is conformal near $\bar{z}_j$ and, by choosing a small enough neighborhood of $\bar{z}_j$, the regions labelled 1, 2, and 3 are approximately equal sectors rotated by a factor that takes account of $\Pi_{\ell \neq j}(\bar{z}_j - \bar{z}_\ell)$. This is illustrated in Figure 5.1.

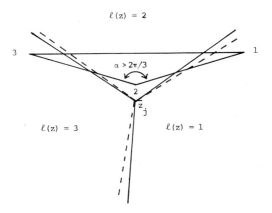

Figure 5.1

By an argument that parallels the proof of Proposition
4.1, for a small enough $\mu(T)$, there is a circuit around the
boundary of $N_\rho(\bar{z}_j)$ with exactly one edge labelled $(1,2)$
and no edge labelled $(2,1)$ in counterclockwise order.
Hence, by the Combinatorial Stokes' Theorem of [1], p. 157,
there exists one more triangle labelled $(1,2,3)$ in cyclic
order than triangles labelled $(1,3,2)$ inside $N_\rho(\bar{z}_j)$.

However, Figure 5.1 shows that it is impossible to have a
triangle labelled $(1,3,2)$ near $\bar{z}_j$ when $\alpha(T) < 2\pi/3$.

Corollary 5.2: For $0 < \varepsilon < \varepsilon_o$,

$$\lim_{k \to \infty} z_{jk}(\varepsilon) = \bar{z}_j(\varepsilon) \quad ,$$

where $\{\bar{z}_1(\varepsilon),\ldots,\bar{z}_n(\varepsilon)\}$ are the distinct roots of $f(z) + \varepsilon$.

Proof: For the special triangulations introduced in Section 3, $\alpha(T) = \pi/2 < 2\pi/3$. Hence, the sequences $(z_{jk}(\varepsilon))$ induced by $f(z) + \varepsilon$ for $0 < \varepsilon < \varepsilon_o$ pass through the unique completely labelled triangles near $\bar{z}_j(\varepsilon)$ at high enough d and converge to $\bar{z}_j(\varepsilon)$ as $k \to \infty$. ∎

Finally, having chosen a fixed order for the starts on $\partial Q_m(\bar{z};h)$, independent of f and ε , let

$$z_j^- = \lim_{\varepsilon \to 0+} \bar{z}_j(\varepsilon) \ .$$

These limits exist and are a complete set of roots of f because the roots of a polynomial are continuous functions of the coefficients.

The proof of Theorem 5.1 would now be completed by the equations:

$$\bar{z}_j = \lim_{\varepsilon \to 0+} \bar{z}_j(\varepsilon)$$

$$= \lim_{\varepsilon \to 0+} \lim_{k \to \infty} z_{jk}(\varepsilon)$$

$$= \lim_{k \to \infty} \lim_{\varepsilon \to 0+} z_{jk}(\varepsilon)$$

$$= \lim_{k \to \infty} z_{jk} \ .$$

All of these equations have been established except for the validity of the interchange of the double limit. This seems

quite difficult to establish directly. However, we can prove
this equation indirectly.

For each j and ε , $0 < \varepsilon < \varepsilon_o$, let z_{jd} (and
$z_{jd}(\varepsilon)$) be the first z_{jk} (and $z_{jk}(\varepsilon)$) that reaches the
level $d = 0,1,\ldots$. These are subsequences that converge
to the same limits. That is:

$$\bar{z}_j = \lim_{\varepsilon \to 0+} \bar{z}_j(\varepsilon)$$

$$= \lim_{\varepsilon \to 0+} \lim_{d \to \infty} z_{jd}(\varepsilon)$$

and

$$\lim_{d \to \infty} z_{jd} = \lim_{k \to \infty} z_{jk} \quad .$$

Moreover, by Lemma 5.1, $\displaystyle\lim_{\varepsilon \to 0+} z_{jd}(\varepsilon) = z_{jd}$ and so

$$\lim_{d \to \infty} \lim_{\varepsilon \to 0+} z_{jd}(\varepsilon) = \lim_{d \to \infty} z_{jd} = \lim_{k \to \infty} z_{jk} \quad .$$

Therefore, Theorem 5.1 is proved if the interchange of limits
is valid for the modified sequences $z_{jd}(\varepsilon)$. However, for
these sequences, the convergence of $\displaystyle\lim_{d \to \infty} z_{jd}(\varepsilon)$ is uniform
in ε over $0 < \varepsilon < \varepsilon_o$ since the sequence reaches level d
in d steps no matter what ε is.   ∎

Research supported by the National Science Foundation under Grant NSF–MPS 72–04983 A 02.

REFERENCES

[1] Kuhn, H. W. "A New Proof of the Fundamental Theorem of Algebra," in Mathematical Programming Study 1, Ed. M. Balinski, North-Holland (1974), 148-158.

[2] Lemke, C. E., "Bimatrix Equilibrium Points and Mathematical Programming," Management Science 11 (1964-5), 681-689.

[3] Lemke, C. E., and Howson, J. T., Jr., "Equilibrium Points of Bimatrix Games," Journal of SIAM 12 (1964), 412-423.

[4] Scarf, H., "The Approximation of Fixed Points of a Continuous Mapping," SIAM Journal of Applied Mathematics 15 (1967), 1328-1343.

[5] Kuhn, H. W., "Simplicial Approximation of Fixed Points," Proceedings, National Academy of Science 61 (1968), 1238-1242.

[6] Merrill, O. H., "Applications and Extensions of an Algorithm that Computes Fixed Points of Certain Upper Semi-Continuous Point-to-Set Mappings," University of Michigan, Ph.D. Thesis, 1972.

[7] Kuhn, H. W., and MacKinnon, J. G., "Sandwich Method for Finding Fixed Points," Journal Opt. Theory and Appl. 17 (1975), 189-204.

[8] Eaves, B. C., "Homotopies for Computation of Fixed Points," Mathematical Programming 3 (1972), 1-22.

[9] Eaves, B. C., and Saigal, R., "Homotopies for Computation of Fixed Points on Unbounded Regions," Mathematical Programming 3 (1972), 225-237.

A New Simplicial Approximation Algorithm With Restarts:

Relations Between Convergence and Labelings

M. L. Fisher, F. J. Gould, and J. W. Tolle

ABSTRACT

    A simplicial approximation algorithm is presented which is applicable to the fixed point problem in R^n or in R^n_+ , the nonlinear complementarity problem, and the problem of solving a system of nonlinear equations in R^n . The algorithm employs a variable initial point, admits a restart procedure, and uses a broad class of matrix labels. This generality yields convergence and existence results under rather weak assumptions. Examples are presented which emphasize relations between labelings and convergence, and illustrate the potential for solving nonlinear equations when Newton's method fails.

1. Introduction

Let $g(x)$ be a map from R^n to R^n . Let $f(x) =$
$x - g(x)$. Then the problem of computing a fixed point of
$g(x)$ is equivalent to the problem of computing a solution to
the system of equations, $f(x) = 0$. Let $g(x)$ be a map
from R_+^n to R_+^n . Let $f(x) = x - g(x)$. Then the problem of computing a fixed point of $g(x)$ is equivalent to the
problem of solving the nonlinear complementarity problem

NLCP: $x \geq 0$, $f(x) \geq 0$, $<x, f(x)> = 0$.

In this paper, we extend the theoretic characterization
of a simplicial approximation algorithm first presented by
the authors in [2] and [3]. The same basic algorithm is applicable both to the problem of finding a zero and to finding
a NLCP solution. By the above remarks, then, the application
to fixed point computation is obvious. More generally, the
algorithm has been extended in [3] to solve a problem in which
some components of a map $f(x)$ are required to satisfy a
complementarity condition and the other components are required to be zero. This more general problem therefore includes as special cases the nonlinear complementarity problem and the problem of solving a system of nonlinear equations. It will be convenient to address the remarks in this
paper specifically to the latter problem of solving nonlinear
equations.

Two important features of the work to be presented are:
(i) the algorithm allows for a variable starting point; and
(ii) the labeling functions have considerable structure.

These two features lead to a new class of convergence
results under rather weak hypotheses. Constructively, then,
new existence theorems are also implied. Examples will be
used to illustrate these comments.

The power of the method for solving a system of nonlinear equations will be illustrated with an example in R^2
where Newton's method fails whenever $|x_1| > 1$, but where
the simplicial algorithm, initiated from essentially any
starting point, converges.

2.   The Algorithm

We present an algorithm for computing a solution, or an
approximate solution, to $f(x) = 0$ where f is a continuous
map from R^n to R^n . Our use of the term algorithm encom-
passes explicitly both a triangulation and a labeling function. The reader should be familiar with the technique of
complementary pivoting on a labeled triangulation. The space
$R^n \times [0,1]$ is triangulated in a manner based on Kuhn's triangulation of a unit cube [4]. A similar triangulation was
employed by Merrill [5] in his dissertation on the computation of fixed point for point to set maps. The symbol δ will
be used to denote the mesh, or grid size. The labeling functions to be introduced give the structure for new results.
They are such that, given any w in R^n , there will be a
unique starting simplex in $R^n \times \{0\}$, with this simplex resid-
ing near w . In other words, the freedom of choice, for

the point w , allows for initiation of the algorithm at any location. As usual, the algorithm cannot cycle, and hence is either unbounded or finite. In the latter case, termination occurs in R^n X $\{1\}$, with the terminal simplex providing an approximate solution to the system of equations. Then sequentially, employing finer and finer grids, the algorithm may be restarted near each previous approximation, yielding seriatim more accurate output.

The labeling functions employ different rules for vertices in R^n X $\{0\}$ and R^n X $\{1\}$. Both rules will depend upon a given point w in R^n . The symbol L_o^w will denote the rule in R^n X $\{0\}$ and L_1^w the rule in R^n X $\{1\}$. The symbol L^w is used to represent the two rules, L_o^w , L_1^w .

First consider two possibilities for L_o^w . Either can be used. Heretofore, w is some chosen point in R^n . Let $\begin{bmatrix} x \\ 0 \end{bmatrix}$ denote a vertex in R^n X $\{0\}$. Abbreviate $L_o^w \left(\begin{bmatrix} x \\ 0 \end{bmatrix} \right)$ as $L_o^w (x)$.

$$L_o^w (x) = \begin{cases} n + 1 & \text{if } x-w > 0 \\\\ i & \text{if } x_i -w_i \le 0 \text{ and } x_j -w_j > 0 \text{ for } j < i . \end{cases} \qquad (1)$$

$$L_o^w (x) = \begin{cases} n + 1 & \text{if } x-w > 0 \\ i & \text{if } x-w \not> 0 \text{ and } x_i -w_i \le x_j -w_j , \text{ all } j \\ (\text{if } i \text{ is not unique, select the least such index}) \end{cases} . (2)$$

It has been shown [2] that with either of these labelings there is a unique $(n+2)$ – almost completely labeled starting simplex in $R^n \times \{0\}$, and this simplex is near the point w .

Now consider the labelings in $R^n \times \{1\}$. These will be called matrix labelings. Abbreviate $L_1^w \begin{matrix} x \\ 1 \end{matrix}$ as $L_1^w(x)$.

Let C, D, and E be given nonsingular $n \times n$ matrices. For the moment we postpone the key question of how to choose these matrices. Define $h^w(x) = Df(C[w + E^{-1}(x-w)])$. Consider the two possible rules

$$
L_1^w (x) = \begin{cases} n + 1 & \text{if } h^w(x) > 0 \\[2mm] i & \text{if } h_i^w(x) \le 0 \text{ and } h_j^w(x) > 0 \text{ for } j < 1. \end{cases}
\tag{3}
$$

$$
L_1^w (x) = \begin{cases} n + 1 & \text{if } h^w(x) > 0 \\[2mm] i & \text{if } h^w(x) \not\vert\; 0 \text{ and } h_i^w(x) \le h_j^w(x), \text{ all } j. \end{cases}
\tag{4}
$$

(if i is not unique, select the least such index)

Again, either of these will be acceptable. Rules (1) and (3) are termed the _natural_ labelings, rules (2) and (4) the _min_ labelings.

It is possible to assign further structure to the labelings by adjoining to the above rules permutations on designated subsets of the indices $\{1, \ldots, n\}$. This additional structure has proved important in application to nonlinear programming (see [3]), but will not be needed in the present discussion.

3. Termination and Existence Results

Given w in R^n we employ the concept of a _separating set_ and a _band_. The idea of a separating set is purely topological. The concept of a band is labeling-dependent.

Definition 1. Suppose w, A, and B are such that A is a bounded and open set in R^n , w $\in$ A , and $\partial A = B$, the boundary of A . Then B is said to _separate_ w _from infinity_. Whenever we refer to the pair (w,B) , it is understood that B separates w from infinity.

Definition 2. Suppose (w,B) and L^w are such that for each x in B there is a label $\ell \in \{1,\ldots,n+1\}, \ell$ depending upon x , such that this label cannot occur in some neighborhood of x.[1] Then the triple (x,B,L^w) is said to be a _band_.

Given the pair (w,B) , let us now define the set

$$\hat{A} = \left\{ u \mid u = C[w + E^{-1}(x - w)], \ x \in A \right\} .$$

It has been shown elsewhere ([2], [3]) that if (w,B,L^w) is a band then if δ is sufficiently small the algorithm must terminate in A X {1} . The terminal simplex provides an approximate solution to $h^w(x) = 0$. Note that x^* in A

---

[1] In other words, there is a neighborhood of x , say N(x) , such that, for every y in N(x) , $L_o^w(y) \neq \ell$, $L_1^w(y) \neq \ell$.

is a zero of h^W , if and only if, u^* in $\hat{A}$ is a zero of
f , where u^* is given by $C[w + E^{-1}(x^* - w)]$. Under the
above transformation, then, x^* is mapped into a point in
which is an approximate solution to $f(x) = 0$. By repeated-
ly restarting from the same initial point w , with succes-
sively finer grids, it is seen that if (w,B,L^W) is a band
about w , then $h^W(x)$ has a computable zero in the set $\hat{A}$,
and consequently f has a zero, also computable, in $\hat{A}$.
From these remarks, it follows that the existence of solutions and algorithmic convergence may be proved by giving hypotheses which assure that (w,B,L^W) is a band.

The importance of the band concept is illustrated in
Figure 1 on the next page. In this illustration, B separa-
rates w from infinity. Take L^W to be the natural labeling [rules (1) and (3)], with C, D, and E chosen as the
identity matrix. It can be seen that every x in B except
for the single point $(0,1)$ complies with the stipulations
of Definition 2. Hence, (w,B,L^W) is not a band. It can be
seen that the eccentric point $(0,1)$ permits the path of
generated simplices to escape and the algorithm is unbounded.
Thus, the behavior at a single point can be critical. If w
is changed, then, of course, the labeling function L^W is
also changed. It is worth noting that if w is placed, for
example, in the northwest sector of the disc in Figure 1,
then the algorithm converges, although there is still not a
band. There is, of course, a possibility that a more judi-
cious choice of the matrices C, D, and E will lead to convergence for a more broad set of initial points, w . This

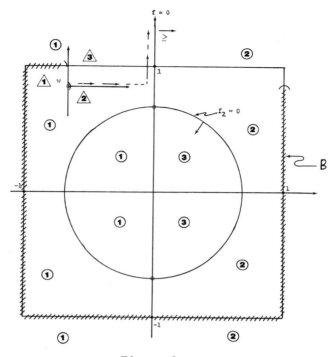

Figure 1

△ indicates labels in regions of $R^2 \times \{0\}$.

○ indicates labels in regions of $R^2 \times \{1\}$.

1. If $x \in B$ with $x_1 > 0$, there is a neighborhood of x in which the label 1 is forbidden.

2. If $x \in B$ with $x_1 < 0$, and $x_2 > -1$, or $x_1 < w_1$ and $x_2 = -1$, then there is a neighborhood of x in which the label 2 is forbidden.

3. ///// denotes points x in B having a neighborhood in which the label 3 is forbidden.

4. The path of simplices: ➝ — — ➝

5. $(0,1)$ has label 1. The labels 1,2,3 appear in every neighborhood of $(0,1)$. Hence (w,B,L^w) is not a band.

matter will be pursued in Section 4.

The following three theorems give hypotheses under which (w, B, L^w) is a band. From the remarks above, it is clear that they can be interpreted as existence and convergence results.

Given a pair (w, B) , and a nonsingular matrix E , it is convenient to define the sets

$$\hat{A} = \{u | u = w + E(x - w) , x \in A\}$$

$$\hat{B} = \{u | u = w + E(x - w) , x \in B\} .$$

Note that since B separates w from infinity, it is also true that $\tilde{B}$ separates w from infinity.

Theorem 1. Let L_o^w be given by (1) or (2), and L_1^w by (3) or (4), where the pair (w, B) and the nonsingular matrices C, D, and E are such that at each x in B at least one of the following is true:[1]

$$\max_{i} E(x - w)_i \ Df(Cx)_i > 0 \qquad (5)$$

$$E(x - w) \not\gtrless 0 , \quad Df(Cx) \not\gtrless 0 . \qquad (6)$$

Then, $(w, \tilde{B}, L^w)$ is a band and f has a zero in CA .

---

[1] The notation $E(x - w)_i$ denotes the ith component of the vector $E(x - w)$, etc.

Proof. The hypotheses imply that at each u in $\tilde{B}$ either

$$\max_i \, (u_i - w_i) \, h_i^w \, (u) > 0$$

or

$$u - w \not\gtrless 0 \, , \quad h^w \, (u) \not\gtrless 0 \, .$$

If $u - w \not\gtrless 0$, $h^w \, (u) \not\gtrless 0$, then there exist i and j for which $u_i - w_i < 0$ and $h_j^w \, (u) < 0$, which implies the label $n + 1$ is excluded in a neighborhood about u . Similarly, if $\max_i \, (u_i - w_i) \, h_i^w \, (u) > 0$, then, for some j , $(u_j - w_j) \, h_j^w \, (u) > 0$. If both terms in the product are positive, the label j is excluded in a neighborhood of u , while if both are negative, $n + 1$ is excluded. Consequently, by definition, $(w,\tilde{B},L^w)$ is a band. This means that h^w has a zero in $\tilde{A}$ and then, from the definitions, f has a zero in CA .

Theorem 1 applies to either the min or the natural labeling. By restricting attention to each of these, separately, the existence of a band may be established under weaker hypotheses as follows. Let e in R^n denote the vector $e_i = 1, 1 \leq i \leq n$.

Theorem 2. Let L^w be the min labeling, where the pair (w,B) and the nonsingular matrices C, D, E are such that at each x in B , the following conditions are satisfied:

(i)   If $E(x-w) \geq 0$, $Df(Cx) \geq 0$, then $\langle E(x-w), Df(Cx) \rangle = 0$.

(ii)   If $E(x-w) \geq 0$, $Df(Cx) \not\geq 0$, then there is a y in R^n satisfying

$$
\begin{cases}
\langle Df(Cx) , E(x - w - y) \rangle > 0 \\
\quad\quad \langle e , E(x - w - y) \rangle \geq 0 \\
\quad\quad\quad\quad\quad\quad Ey \geq 0 \quad .
\end{cases}
\tag{7}
$$

(iii)   If $E(x-w) \not\geq 0$, $Df(Cx) \geq 0$, then there is a y in R^n satisfying

$$
\begin{cases}
\langle E(x - w) , Df(Cx) - y \rangle > 0 \\
\quad\quad \langle e , Df(Cx) - y \rangle \geq 0 \\
\quad\quad\quad\quad\quad\quad y \geq 0 \quad .
\end{cases}
\tag{8}
$$

Then $(w,\tilde{B},L)$ is a band and f has a zero in CA .

Proof. Note that (i) implies that systems (7) and (8) are trivially satisfied by $y = 0$. Also, if x in B is such that $E(x - w) \not\geq 0$, $Df(Cx) \not\geq 0$, then the first and third conditions of system (8) can be tricially satisfied by making an appropriate component of y suitably large and setting the other components to zero. Note also that if $x \in B$ is in case (ii), then the systems (7) and (8) are either trivially satisfied as in case (i) or else (8) cannot hold. A similar comment applied to case (iii). From these remarks, the hypotheses of the theorem are equivalent to asserting that at each x in B there is a y such that at least one of the following two systems is satisfied:

$$
\begin{cases}
\langle Df(Cx) \ , \ E(x - w - y)\rangle \ > \ 0 \\
\qquad \langle e \ , \ E(x - w - y)\rangle \ \geq \ 0 \quad \text{if} \quad E(x - w) \ \geq \ 0 \qquad (9) \\
\qquad\qquad\qquad Ey \ \geq \ 0
\end{cases}
$$

$$
\begin{cases}
\langle E(x - w) \ , \ Df(Cx) - y\rangle \ > \ 0 \\
\qquad \langle e \ , \ Df(Cx) - y\rangle \ \geq \ 0 \quad \text{if} \quad Df(Cx) \ \geq \ 0 \qquad (10) \\
\qquad\qquad\qquad y \ \geq \ 0 \quad .
\end{cases}
$$

It follows then from the definitions that at every u in $\tilde{B}$ there is a nonnegative $\hat{y}$ in R^n such that at least one of the following systems is satisfied:

$$
\begin{cases}
\langle u - w - y \ , \ h^w(u)\rangle \ > \ 0 \\
\qquad \langle e \ , \ u - w - y\rangle \ \geq \ 0 \quad \text{if} \quad u - w \ \geq \ 0
\end{cases} \qquad (11)
$$

$$
\begin{cases}
\langle u - w \ , \ h^w(u) - y\rangle \ > \ 0 \\
\qquad \langle e \ , \ h^w(u) - y\rangle \ > \ 0 \quad \text{if} \quad h^w(u) \ \geq \ 0 \quad .
\end{cases} \qquad (12)
$$

It has been shown in [2] that this implies $(w,\tilde{B},L^w)$ is a band, which gives the first conclusion of the theorem. The second conclusion is immediate from the definitions.

The next result will give similar conditions for the existence of a band for a natural labeling. For brevity, the proof will be omitted. It is based on the same transformations employed in Theorems 1 and 2 and the arguments in [3].

Theorem 3. Let L^W be the natural labeling, where the pair (w,B) and the nonsingular matrices C,D,E are such that at each x in B the following conditions are satisfied:

(i)   If $E(x-w) \geq 0$ and $Df(Cx) \geq 0$, then $\langle E(x-w), Df(Cx) \rangle \neq 0$.

(ii) If $E(x-w) \geq 0$ and $Df(Cx) \not\geq 0$, then there exists y satisfying

$$\langle Df(Cx) , E(x - w - y) \rangle > 0$$

$$E(x - w - y)_i \geq 0 , i \geq \max \{j \mid E(x - w)_j > 0\}$$

$$Ey \geq 0 .$$

(iii) If $E(x-w) \not\geq 0$ and $Df(Cx) \geq 0$, then there exists $y \geq 0$ satisfying

$$\langle E(x - w) , Df(Cx) - y \rangle > 0$$

$$Df(Cx)_i - y_i \geq 0 , i \geq \max \{j \mid Df(Cx)_j > 0\} .$$

Then $(w,\tilde{B},L^W)$ is a band and f has a zero in CA .

It is worth remarking that the hypotheses of the above theorem are equivalent to asserting that at each x in B there is a y such that at least one of the following two systems is satisfied:

$$
\left\{
\begin{array}{l}
\langle\, Df(Cx)\ ,\ E(x - w - y)\rangle\ >\ 0 \\[6pt]
\text{If}\quad E(x - w)\ \geq\ 0\ ,\ \text{then} \\[6pt]
\qquad E(x - w - y)_i\ \geq\ 0\quad \text{for all} \\[6pt]
\qquad i\ \geq\ \max\ \{j\,|\,E(x - w)_j\ >\ 0\} \\[6pt]
\quad Ey\ \geq\ 0\quad .
\end{array}
\right. \tag{13}
$$

$$
\left\{
\begin{array}{l}
\langle\, E(x - w)\ ,\ Df(Cx) - y\rangle\ >\ 0 \\[6pt]
\text{If}\quad Df(Cx)\ \geq\ 0\ ,\ \text{then} \\[6pt]
\qquad Df(Cx)_i\ -\ y_i\ \geq\ 0\quad \text{for all} \\[6pt]
\qquad i\ \geq\ \max\ \{j\,|\,Df(Cx)_j\ >\ 0\} \\[6pt]
\quad y\ \geq\ 0\quad .
\end{array}
\right. \tag{14}
$$

The reader can verify that the hypotheses of Theorem 1 imply those of both Theorems 2 and 3. Also we state without elaboration that a number of known existence results can be obtained from the above theorems with an appropriate choice for the matrices D, C, and E .

4. Implementation

It is now evident that a key problem in the application of this theory is to find matrices C, D, and E , and initial points w , such that the hypotheses of one or more of the above theorems hold, for this would assure algorithmic convergence. Given $w \in R^n$, we let $J_f(w)$ denote the Jacobian of f at w . It has been shown in [2] that if x^* is a solution to $f(x) = 0$ at which f is continuously

differentiable and at which $J_f(x^*)$ is nonsingular, then
there is a bounded open set A containing x^* such that for
every $w \in A$ the hypotheses of Theorem 1 are satisfied by
taking $B = \partial A$, $C = E = I$, and $D = J_f(w)^{-1}$. This implies
that if the initial w is suitably close to x^* , and the
initial δ suitably small, then the algorithm converges with
restarts.

A potential computational strength of this method lies
in the possibility of convergence in cases where classical
methods fail. This would imply that for such cases, at least
in low dimension, the method presented in this paper may be
worthwhile.

An illustration of this possibility is provided by the
following example in R^2 . These functions were presented
by Eaves in [1].

$$f_1(x) = \max \begin{cases} x_1 + x_2 - 2 \\ \\ -2x_1 + x_2 + 1 \end{cases} \qquad f_2(x) = \max \begin{cases} -x_1 + x_2 - 2 \\ \\ 2x_1 + x_2 + 1 \end{cases}$$

The following facts can be verified. The point $(0,-1)$ is a
solution to $f(x) = 0$. Newton's method fails whenever the
starting point is such that $|x_1| > 1$. It converges when
$|x_1| > 1$. The procedure described herein, taking $C = E$
$= I$, $D = J_f(x)^{-1}$, converges when initiated from any $w \in R^2$
where f is differentiable (any w such that $|w_1| \neq 1$).
In this example, the importance of using the Jacobian is

reflected in the fact that if D is chosen to be the identity matrix, then the algorithm is unbounded no matter where
it started.

The work of the first two authors was supported in part by NSF Grant GS–42010 and in part by ONR Contract N–00014–A–0285–0019. The second author acknowledges the considerable assistance of Victor Neumark in the support of this research.

REFERENCES

[1] Eaves, B. C., "Solving Regular Piecewise Linear Convex
     Equations", to appear in Mathematical Programming.

[2] Fisher, M. L., Gould, F. J., and Tolle, J. W., "A Simpli-
     cial Approximation Algorithm for Solving Systems of Non-
     linear Equations", Center for Mathematical Studies in
     Business and Economics Report No. 7421, Graduate School
     of Business, University of Chicago, May 1974.

[3] Fisher, M. L., Gould, F. J., and Tolle, J. W., "A Simpli-
     cial Algorithm for the Mixed Nonlinear Complementarity
     Problem with Application to Convex Programming", Center
     for Mathematical Studies in Business and Economics Report
     No. 7424, Graduate School of Business, University of
     Chicago, May 1974.

[4] Kuhn, H. W., "Some Combinatorial Lemmas in Topology",
     IBM J. Res. Develop., Vol. 4, 1960, pp. 508-524.

Complementary Pivot Theory and Markovian Decision Chains

B. Curtis Eaves

ABSTRACT

Techniques of complementary pivot theory are used for solving the fixed point problem

$$y = \max_{\delta} P_{\delta} y + R_{\delta}$$

under new conditions.

0. Notation

Let R^{mxn} denote the set of mxn real matrices and let $R^n = R^{nx1}$. If $A \in R^{mxn}$ and $\beta \subset \{1,\ldots,m\}$ or $\beta \subset \{1,\ldots,n\}$, then by $A_{\beta \cdot} \in R^{|\beta|xn}$ or $A_{\cdot \beta} \subset R^{mx|\beta|}$ we denote the corresponding submatrix of rows or of columns of A. If A is in R^{1xn} or R^n we often omit the dot. By $A \geq 0$ or $A > 0$ we denote that every element is non-negative or positive, respectively.

1. Introduction

Given the nonempty subsets Δ_i of $R^{1x(n+1)}$ for $i \in S = \{1,\ldots,n\}$ let $\delta \in \Delta$ index the matrices (P,R) in $\Delta_1 x \ldots x \Delta_n$, that is, those matrices in $R^{nx(n+1)}$ with $(P,R)_{i \cdot} \in \Delta_i$ and $P_{i \cdot} \in R^{1xn}$. Let us define the affine functions $L_\delta : R^n \rightarrow R^n$ and the function $L : R^n \rightarrow R^n$ by

$$L_\delta(y) = P_\delta y + R_\delta$$

$$L(y) = \sup_{\delta \in \Delta} P_\delta y + R_\delta = \sup_{\delta \in \Delta} L_\delta(y)$$

We assume throughout the entire paper that L is finite valued. The components of L, being the supremum of affine functions, are convex. The central thrust of this paper is to compute, finitely quick, fixed points $y = L(y)$ of L

under new global conditions. Perhaps our main result, given in §7 , is that if the set $\{y : |L_\delta(y) \leq y \leq L(y)\}$ is bounded for some $\delta \in \Delta$ with $\det(I - P_\delta) \neq 0$ then the algorithm will compute a fixed point of L .

To sketch the algorithm we begin by identifying δ_o together with an attendant assumption and by defining d_δ for all δ . Let δ_o be a distinguished element of Δ ; we call δ_o the primary policy. Throughout the entire paper we assume that $\det(I - P_{\delta_o}) \neq 0$ where $I \in R^{n \times n}$ is the identity matrix; we call this assumption "primary regularity". For $\delta \in \Delta$ define $d_\delta \in R^n$ by $(d_\delta)_i = 0$ or 1 corresponding to whether or not $\delta(i) = \delta_o(i)$, that is whether or not $(P_\delta, R_\delta)_{i\cdot} = (P_{\delta_o}, R_{\delta_o})_{i\cdot}$.

For $\delta \in \Delta$ and $0 \leq z \in R^1$ we define the family of affine functions $L_\delta^z : R^n \rightarrow R^n$ by

$$L_\delta^z(y) = P_\delta y + R_\delta - d_\delta z$$
$$= L_\delta(y) - d_\delta z$$

and the family of functions $L^z : R^n \rightarrow R^n \cup \{\infty\}$ by

$$L^z(y) = \sup_{\delta \in \Delta} L_\delta^z(y)$$

Assuming Δ is finite, the algorithm begins by solving the simplified fixed point problem $L^z(y) = y$ for some $z > 0$. Then with a sequence of pivots each motivated by a "complementary pivot logic" as in Lemke's algorithm [16] for quadratic programming and matrix games and by a "policy improvement logic" as in Howard's algorithm [13] for dynamic

programming, a piecewise linear path $(y(t), z(t))$ for $t \in R^1$
is generated such that

$$L^{z(t)}(y(t)) = y(t)$$

Either $z(t) = 0$ for some t or $y(t) \to \infty$ as $t \to +\infty$ or
both. Arguing with and about this path we obtain the results
regarding fixed points of L .

The algorithm is described in §5 ; preliminaries for
the description are given in §§3 and 4 . A precise
statement of the algorithm output, namely the path $(y(t)$,
$z(t))$, is given in §6 . In the next § , §2 , a relationship of the fixed point problem $L(y) = y$ to dynamic
programming is discussed; the framework is used in §§8 and
9 .

2.   Relation to Dynamic Programming

In the context of dynamic programming one can think of
a process that transits at times $t = 0,1,\ldots$ among the
states $s \in S$ or to the "stopped" state. One thinks of
$(p,r) \in \Delta_i$ as an available alternative when the process is
in state i , of r as the immediate return for being in
state i and using alternatives (p,r) , of p_j where
$p = (p_1,\ldots,p_n)$ as the probability that the process transits
to state j , and of $1 - \sum p_i$ as the probability that the
process permanently stops.

For example, assume that $P_\eta \geq 0$ and that $P_\eta^k \to 0$ as
$k \to +\infty$ for $\eta \in \Delta$. If one uses the stationary policy δ ,

that is, if one selects alternative $(P_\delta, R_\delta)_i$. when the process is in state i , then it follows that the return, that is the expected total return, is

$$y_s = ((\sum_{j=0}^{\infty} P_\delta^j)R_\delta)_s = ((I-P_\delta)^{-1}R_\delta)_s$$

where s is the initial state. If $y = L(y) = L_\delta(y)$ then it can be shown that δ maximizes the return $y = (y_1, \ldots, y_n)$ among all policies that are not clairvoyant, see Shapley [19], Derman [7], Howard [13], Blackwell [3], and Veinott [21].

Our assumptions and data are often inappropriate for this dynamic programming interpretation or for others (e.g., we do not always assume $P_\delta \geq 0$), however, we continue to use the terminology and speak of states, alternatives, policies $\delta \in \Delta$, etc.

Though a few of our results pertain directly to dynamic programming, our principal contribution in this vein is likely to be the introduction of new arguments.

3. Restating $L^z(y) = y$

To precisely describe the algorithm we being by stating the system $L^z(y) = y$ in another form.

Solving $L^z(y) = y$ is equivalent to solving

$$\max_\delta (P_\delta - I) y + R_\delta - d_\delta z = 0$$

or to solving

$$-(I - P_\delta) y - d_\delta z \leq -R_\delta \qquad \delta \in \Delta$$

with equality for some δ .

For our next statement we need for $\Delta_1 \cup \Delta_2 \ldots \cup \Delta_n$ to be finite and ordered. Let us assume that $|\Delta_i| = m_i$ and that

$$\Delta_i = \{(p_j, r_j) : |j \in \alpha_i\}$$

where

$$\alpha_i = \{m_1 + \ldots + m_{i-1} + 1, \ldots, m_1 + \ldots + m_i\}$$

for $i \in S$. We can view Δ as all subsets of $\alpha_1 \cup \ldots \cup \alpha_n$ which contains exactly one element from each α_i for $i \in S$. Let δ_o , the primary policy, be that policy which uses the first alternative at each state, that is,

$$\delta_o = \{\delta_o(1) < \delta_o(2) < \ldots < \delta_o(n)\}$$

$$= \{1, m_1 + 1, \ldots, m_1 + \ldots + m_{n-1} + 1\} .$$

In addition let $\alpha_{n+1} = \{m+1, \ldots, m+n\}$ and $\alpha = \alpha_1 \cup \ldots \cup \alpha_{n+1} \cup \{m+n+1\} = \{1, \ldots, m+n+1\}$.

Let $M \in R^{m \times n}$ and $q \in R^m$ be the matrix and vector

$$\begin{pmatrix} e_1 - p_1 \\ \cdot \\ \cdot \\ \cdot \\ \cdot \\ \cdot \\ \cdot \\ \cdot \\ e_m - p_m \end{pmatrix} - \begin{pmatrix} r_1 \\ \cdot \\ \cdot \\ \cdot \\ \cdot \\ \cdot \\ \cdot \\ r_m \end{pmatrix}$$

where $e_i = (0,\ldots,0,1,0,\ldots,0)$ with the 1 in the jth

position if $i \in \alpha_j$. Consistently with our previous usage

let $d = (d_1,\ldots,d_m) \in R^m$ be defined by

$$d_i = \begin{cases} 1 & i \notin \delta_o \\ \\ 0 & i \in \delta_o \end{cases}$$

Let $x \in R^m$, $y \in R^n$, and $z \in R^1$.

Consider the system

$$Ix - My - dz = q$$

(*)

$$0 \nleq x_{\alpha_i} \geq 0 \quad i \in S \quad z \geq 0$$

where $0 \nleq x_{\alpha_i} \geq 0$ denotes that the components of x_{α_i} are

all nonnegative but not all positive and $I \in R^{m \times m}$ is the

identity.

Lemma 1: If (x,y,z) solves (*) then there is a

policy δ with $x_\delta = 0$ and $L_\delta^z(y) = L^z(y) = y$. Converse-

ly, if $L_\delta^z(y) = L^z(y) = y$, then there is an x with $x_\delta = 0$

such that (x,y,z) solves (*) ☒

Note that $(I,-M,-d) \in R^{mx(m+n+1)}$. If $\beta \subset \alpha$ then

by $\sim\beta$ we denote $\alpha \sim \beta$. By $|\beta|$ we denote the size of

a set β .

4.   Bases and Adjacency

We now define the notion of a basis index. Due to the fact that y is not required to be nonnegative our definition is not the usual one. In addition we have no use for a basis which is "degenerate", hence, in our definition we include the requirement of nondegeneracy.

We say that $A \in R^{kx(m+1)}$ is lexico positive if each row is nonzero and if the first nonzero element of each row is positive.

Definition: If $\beta \subset \alpha$ and if

1. $|\beta| = m$

2. $\beta \supset \alpha_{n+1}$

3. $\beta \not\supset \alpha_i$ for all $i = 1, \ldots, n$

4. $B = (I, -M, -d)^{-1}_{\cdot\beta}$ exists

5. $B(q, I)_{\rho\cdot}$ is lexico positive where $\rho = \beta \sim \alpha_{n+1}$

then we call β a b.i.   (basis index) and (x, y, z) the basic solution of β where

$$(x, y, z)_\beta = Bq$$

$$(x, y, x)_{\sim\beta} = 0$$

Note that such (x, y, z) solves   (*) .

Let β be a b.i. If $m+n+1 \notin \beta$ let $\bar{\beta} = \{m+n+1\}$; otherwise let $\bar{\beta}$ be the unique pair $\{j, k\}$ where $j \neq k$ and

$$\{j, k\} \subset \alpha_i \cap (\sim\beta)$$

for some $i \in S$. If $\beta \neq \gamma$ are b.i.'s we say that they are adjacent if

$$\gamma = \beta \cup \{i\} \sim \{j\}$$

for some $i \in \bar{\beta}$ and $j \in \beta$ or equivalently if

$$\beta = \gamma \cup \{j\} \sim \{i\}$$

for some $j \in \bar{\gamma}$ and $i \in \gamma$.

5. The Algorithm

The algorithm for solving (*) with $z = 0$ proceeds by moving from b.i. to new adjacent b.i. as long as possible. Properly interpreted the algorithm is that of Cottle and Dantzig [4] which in turn is known to be a special case of Lemke [16], see Lemke [15] and Scarf [18].

Initialization: Let β_0 be the b.i. with $\delta_0 \cap \beta_0$ = ϕ and let $\beta_{-1} = \bar{\beta}_0 \sim \delta_0$.

General Step: Assume that the sequence of b.i. $\beta_0, \beta_1, \ldots, \beta_k$ has been generated. If $m+n+1 = \bar{\beta}_k$ terminate. If $m+n+1 \neq \bar{\beta}_k$, determine if there is a b.i. β of form

$$\beta_k \cup \{j\} \sim \{i\}$$

where $j \in \bar{\beta}_k \sim \beta_{k-1}$ and $i \in \beta_k$. If so, let $\beta_{k+1} = \beta$ and repeat the general step; otherwise, terminate.

Note: The tasks of determining β_0 and β_{k+1} given β_k and β_{k-1} are straightforward and consist of running a minimum ratio test as in the simplex method with lexicographic rules for resolving degeneracy; the essential difference here is that α_{n+1} is contained in every b.i. This technique and other features of the algorithm are exhibited in the following example.

An Example: Let $n = 2$, $\Delta_1 = \{(0,0;0), (-4,-7;2)\}$, and $\Delta_2 = \{(0,0;0), (-2,0;1)\}$; hence $\alpha_1 = \{1,2\}$, $\alpha_2 = \{3,4\}$, $\alpha_3 = \{5,6\}$, $\delta_0 = \{1,2\}$, and $m = 4$. The detached coefficient form of (*) is:

1	2	3	4	5	6	7	
1	0	0	0	-1	0	0	0
0	1	0	0	-5	-7	-1	-2
0	0	1	0	0	-1	0	0
0	0	0	1	-2	-1	-1	-1

To determine β_0 and to put the form above into the cannonical form (modulo row permutations) of (*) with respect to β_0 we first block pivot on $\delta_0 \times \alpha_{n+1} = \{1,3\} \times \{5,6\}$ (primary regularity) to get:

1	2	3	4	5	6	7	
-1	0	0	0	1	0	0	0
-5	1	-7	0	0	0	-1	-2
0	0	-1	0	0	1	0	0
-2	0	-1	1	0	0	-1	-1

We see that $(-2,-5,1,-7,0)$ is lexico less than $(-1,-2,0,-1,1)$, therefore, we pivot on $(2,7)$ to obtain the cannonical form of $\beta_0 = \{4,5,6,7\}$:

1	2	3	4	5	6	7	
-1	0	0	0	1	0	0	0
5	-1	7	0	0	0	1	2
0	0	-1	0	0	1	0	0
3	-1	6	1	0	0	0	1

Hence $(x,y,z)(0) = (0,0,0,1,0,0,2)$, $\bar{\beta}_o = \{1,2\}$, $\bar{\beta}_{-1} =$

$\{2\} = \bar{\beta}_o \sim \beta_o$, and $\bar{\beta}_o \sim \beta_{-1} = \{1\}$. Note that by

$(x,y,z)(i)$ we are denoting the basic solution of β_i .

If β_1 exists it is of form $\beta_o \cup \{1\} \sim \{j\}$ with

$j \in \beta_o$. Since $(1,3,-1,6,1)/3$ is lexico less than $(2,5,$

$-1,7,0)/5$ we see that $j = 4$, therefore, we pivot on

$(4,1)$ to get the cannonical form of $\beta_1 = \{1,5,6,7\}$:

1	2	3	4	5	6	7	
0	$-\frac{1}{3}$	2	$\frac{1}{3}$	1	0	0	$\frac{1}{3}$
0	$\frac{2}{3}$	-3	$-\frac{5}{3}$	0	0	1	$\frac{1}{3}$
0	0	-1	0	0	1	0	0
1	$-\frac{1}{3}$	2	$\frac{1}{3}$	0	0	0	$\frac{1}{3}$

Hence $(x,y,z)(1) = \left(\frac{1}{3},0,0,0,\frac{1}{3},0,\frac{1}{3}\right)$, $\bar{\beta}_1 = \{3,4\}$, and

$\bar{\beta}_1 \sim \beta_o = \{3\}$.

If β_2 exists it is of form $\beta_1 \cup \{3\} \sim \{j\}$ with

$j \in \beta_1$. We have only one ratio , $\left(\frac{1}{3},1,-\frac{1}{3},2,\frac{1}{3}\right)\bigg/ 2$, so

$j = 1$, therefore, we pivot on $(4,3)$ to get the cannonical form of $\beta_2 = \{3,5,6,7\}$:

1	2	3	4	5	6	7	
-1	0	0	0	1	0	0	0
$\dfrac{3}{2}$	$\dfrac{1}{6}$	0	$-\dfrac{7}{6}$	0	0	1	$\dfrac{5}{6}$
$\dfrac{1}{2}$	$-\dfrac{1}{6}$	0	$\dfrac{1}{6}$	0	1	0	$\dfrac{1}{6}$
$\dfrac{1}{2}$	$-\dfrac{1}{6}$	1	$\dfrac{1}{6}$	0	0	0	$\dfrac{1}{6}$

Hence $(x,y,z) = \left(0,0,\dfrac{1}{6},0,0,\dfrac{1}{6},\dfrac{5}{6}\right)$, $\bar{\beta}_2 = \{1,2\}$, and $\bar{\beta}_2 \sim \beta_1 = \{2\}$.

If β_3 exists it is of form $\beta_2 \cup \{2\} \sim \{j\}$ with $j \in \beta_2$. Here again we have one ratio , $\left(\dfrac{5}{6},\dfrac{3}{2},\dfrac{1}{6},0,-\dfrac{7}{6}\right)\bigg/\left(\dfrac{5}{6}\right)$, so $j = 7$, therefore, we pivot on $(2,3)$ to get the cannonical form of $\beta_3 = \{2,3,5,6\}$:

1	2	3	4	5	6	7	
-1	0	0	0	1	0	0	0
9	1	0	-7	0	0	6	5
2	0	0	-1	0	1	1	1
2	0	1	-1	0	0	1	1

Hence $(x,y,z)(3) = (0,5,1,0,0,1,0)$, $\bar{\beta}_3 = \{7\}$, and we terminate with the fixed point $y(3) = (0,1)$.

That this example terminates with $z(3) = 0$ follows from Corollary 4. Note that $z(1) = \dfrac{1}{3} < \dfrac{5}{6} = z(2)$. Of course, a serious computer code would manipulate data quite

differently from what we have; for example there is no need
to explicitly store the mxm identity matrix of the cannonical form (the resulting form would be m x (n+2)) .

6. The Algorithm Output

The following theorem, patterned after that of Lemke
[16], describes what the algorithm generates. The proof
rests upon the facts that β_o is unique and adjacent to at
most one b.i. , that any b.i. β is adjacent to at most
two b.i. ($|\bar{\beta}| \leq 2$) , and that if $j \in \bar{\beta}$ and there is no
b.i. of form $\beta \cup \{j\} \sim \{i\}$ for some $i \in \beta$ then one can
generate a ray of solutions to (*) using coordinates
$\beta \cup \{j\}$, see [8,10], Lemke [16], Scarf [18], and Cottle and
Dantzig [4].

Theorem 2. The sequence of b.i. $\beta_o, \ldots, \beta_\ell$ generated
by algorithm is finite, unique, and its elements are distinct. Let $(x,y,z)(i)$ $i = 0, \ldots, \ell$ be the corresponding
basic solutions of (*) . This function $(x,y,x)(\cdot)$ has a
continuous extension to R^1 such that it solves (*) and
hence

$$L^{z(t)}(y(t)) = y(t)$$

for $t \in R^1$. This extension can be chosen and is chosen so
that it is affine on each of the pieces $(-\infty,0], [0,1],[1,2],$
$\ldots,[\ell-1,\ell], [\ell,+\infty)$. On each such piece J there is a δ
such that

$$L_\delta^{z(t)}(y(t)) = L^{z(t)}(y(t))$$

for $t \in J$. On $(-\infty, 0]$ $y(t) = y(o)$ and $z(t) \rightarrow +\infty$ as $t \rightarrow -\infty$. If $z(\ell) > 0$ then $y(t) \rightarrow \infty$ as $t \rightarrow +\infty$ ⊠

We call $(x,y,z)(t)$ for $t \in (-\infty, 0]$ the primary ray and if $y(t) \rightarrow \infty$ as $t \rightarrow +\infty$ we call $(x,y,z)(t)$ for $t \in [\ell,+\infty)$ the secondary ray.

7. A Ray Condition

Given the finite set $\Delta_1 x \ldots x \Delta_n$ and δ_o we describe a condition which prohibits secondary rays, hence, one is guaranteed convergence of the algorithm to a fixed point of L . In the next § an important subclass, wherein the fixed point is unique, is discussed.

Theorem 3: If Δ is finite and the set $\{y : L_{\delta_o}(y) \leq y \leq L(y)\} = \underset{\delta}{\cup} \{y : L_{\delta_o}(y) \leq y \leq L_\delta(y)\}$ is bounded then the algorithm yields $z(\ell) = 0$ and hence $y(\ell)$ is a fixed point of L .

Proof: Throughout the algorithm we have $y(t) = L^{z(t)}(y(t)) \leq L(y(t))$, hence we have

$$L_{\delta_o}(y(t)) \leq y(t) \leq L_\delta(y(t))$$

for $t \in [\ell,+\infty)$ for some $\delta \in \Delta$. In view of the hypothesis, $\{y(t) : t \in [\ell,+\infty)\}$ is bounded, there is no secondary ray, therefore, $m + n + 1 = \bar{\beta}_\ell$, $z(\ell) = 0$, and $y(\ell)$ is a fixed point of L ⊠

Using the continuous version of this argument as in [9] and without assuming $|\Delta| < \infty$ one can show that if $\{y : L_{\delta_o}(y) \leq y \leq L(y)\}$ is bounded then L has a fixed point.

Corollary 4: If Δ is finite, $(P_{\delta_o}, R_{\delta_o}) = 0$, and $P_\delta \leq 0$ for $\delta \in \Delta$ then the algorithm yields a fixed point.

Proof: $\{0 \leq y \leq P_\delta y + R_\delta\}$ is bounded if $P_\delta \leq 0$ ☒

This corollary applies to the example of §5 . Note that $z(t)$ does not decrease in the example, however, as we shall see, $z(t)$ does decrease for the subclass of the next section.

8. Total Regularity

The main result of this section is that if Δ is finite and totally regular then L has a unique fixed point and we can compute it finitely quick.

Definition: If for each $\delta \in \Delta$, $\det (I - P_\delta) \neq 0$ then we say that Δ or $\Delta_1 x \ldots x \Delta_n$ is regular. If $C(\Delta_1) x \ldots x C(\Delta_n)$ is regular we call Δ or $\Delta_1 x \ldots x \Delta_n$ totally regular where $C(\Delta_i)$ is the convex hull of Δ_i for $i \in S$.

Lemma 5: If Δ is totally regular then $\{y : L_{\delta_o}(y) \leq y \leq L_\delta(y)\}$ is bounded for $\delta \in \Delta$.

Proof: Otherwise there is a $y \neq 0$ such that $P_{\delta_o} y \leq y$ and $y \leq P_\delta y$ or that $(I - P_{\delta_o})y \geq 0$ and $(I - P_\delta)y \leq 0$.

On taking a convex combination, row by row, we get that $(I - P)y = 0$ and using total regularity we get $\det(I - P) \neq 0$ or that $y = 0$, that is, we get a contradiction ⊠

If for each $y \in R^n$ there is a $\delta \in \Delta$ such that $L_\delta(y) = L(y)$ when we say that L is achieved.

Lemma 6: If L is achieved and Δ is totally regular, then L has at most one fixed point.

Proof: Let y and y' be fixed points of L , then for some policies δ and γ we have

$$y = P_\delta y + R_\delta \qquad\qquad y' = P_\gamma y' + R_\gamma$$

$$y \geq P_\gamma y + R_\gamma \qquad\qquad y' \geq P_\delta y' + R_\delta$$

or that

$$(I - P_\delta)(y' - y) \geq 0$$

$$(I - P_\gamma)(y' - y) \leq 0 \quad .$$

Again on taking a convex combination, row by row, of P_δ and P_γ we get $(I - P)(y' - y) = 0$ and $\det(I - P) \neq 0$; hence, $y' = y$ ⊠

Theorem 7: If Δ is finite and totally regular then the algorithm terminates with $z(\ell) = 0$ and yields the unique fixed point $y(\ell)$ of L . Furthermore $z(t)$ decreases as $t \in R^1$ increases.

Proof: The first assertion follows from Theorem 3 and Lemma 6. Recalling that $z(t) \to +\infty$ as $t \to -\infty$ and that

$z(\ell) = 0$, suppose we have $z(t) > z(t')$ for some $t < t'$. It follows that there are integers $0 \leq k \leq h \leq \ell - 1$ for which $[z(k), z(k+1)]$ and $[z(h), z(h+1)]$ have a common interior point. Using $\beta_k \neq \beta_{h+1}$ and the last condition in the definition of a b.i. it follows that there are $y \neq y'$, z, and an $\varepsilon \geq 0$ for which

$$y = \max_{\delta \in \Delta} L_\delta^z(y) + [\varepsilon]_\delta$$

$$y' = \max_{\delta \in \Delta} L_\delta^z(y') + [\varepsilon]_\delta$$

where $[\varepsilon] = (\varepsilon, \varepsilon^2, \ldots, \varepsilon^m) \in R^{mx1}$. This contradicts the content of Lemma 6 applied to $\max_\delta L_\delta^z(\cdot) + [\varepsilon]_\delta$ and it follows that $z(t)$ decreases, that is does not strictly increase, as t increases ⊠

Note: If y is a fixed point of L based on $\Delta_1, \ldots,$ Δ_n then y remains a fixed point of L based on $C(\Delta_1)$ $, \ldots, C(\Delta_n)$ where C denotes the convex hull.

If Δ is finite and $|P_\delta|^n \to 0$ (where elements of $|P_\delta| \in R^{nxn}$ are the corresponding absolute values of elements of P_δ) as $n \to \infty$ for each $\delta \in \Delta$ then Δ is totally regular and L has a unique fixed point. This result, to our knowledge, was previously obtained through a value iteration argument based on a generalized Banach contraction mapping result. Here one demonstrates that $L^n(y_o) \to y = L(y)$

as $n \to \infty$, see Shapley [19], Denardo [5], and Veinott [21].
If in addition $P_\delta \geq 0$ one can also show that policy im-

provement and hence linear programming will obtain the fixed
point, see Denardo [5], Howard [13], Veinott [21], D'Epenoux
[6], and Grinold [12]. Other than these special cases, Theorem 3 is new, at least, to our knowledge.

The monotone behavior of $z(t)$ as stated in Theorem 4
leads one to the following two questions: 1) Assuming total
regularity could one solve (*) by using linear programming
and minimizing z ? 2) Does policy improvement (or Newton's
method) actually work under a larger subclass of the totally
regular class then hitherto proved?

Question 1 is easily shown to be false if we are
speaking of the standard simplex method; one would get $z = 0$
but one would probably not obtain $x_{\alpha_i} \ne 0$ in the final so-

lution. If one uses a restricted basis entry and insist on
$x_{\alpha_i} \ne 0$ throughout, then one gets essentially the algorithm

that we propose. With regard to question 2 we note that
proof of convergence will necessarily be quite different from
the customary one based on monotonicity; perhaps one could
adapt our proof of Theorem 4.

The task of finding the value and associated strategies
of Shapley's [19] two-person terminating stochastic game with
perfect information can be formulated as a fixed point problem of form $L(y) = y$; briefly we indicate how this is done.
Player σ selects the alternative $(p,r) \in \Delta_i'$ at states
$i \in \sigma = \{1,\dots,k\}$ and τ selects the alternative $(p,r) \in$
Δ_i' for states $i \in \tau = \{k+1,\dots,n\}$. Assuming $P_\delta \geq 0$ and

$P_\delta^k \to 0$ for $\delta = (\delta_\sigma, \delta_\tau) \in \Delta'$, if the game begins in state

$i \in S$ and the play is according to $\eta = (\eta_\sigma, \eta_\tau) \in \Delta'$ then

τ pays σ the amount

$$\left(\left(\sum_{k=0}^{\infty} P_\eta\right) R_\eta\right)_i$$

Define Δ_i to be

$$\{((p_\sigma, -p_\tau), r) : | (p, r) \in \Delta_i'\}$$

for $i \in \sigma$ and

$$\{((-p_\sigma, p_\tau), -r) : |(p, r) \in \Delta_i'\}$$

for $i \in \tau$. If $L(y) = L_\delta(y) = y$ we have y_i or $-y_i$ as

the value of the game if play begins in state $i \in \sigma$ or

$i \in \tau$ and δ_σ and δ_τ as optimal strategies for σ and

τ respectively where $(\delta_\sigma, \delta_\tau) = \delta$. Note Δ is totally

regular and if Δ is finite our algorithm will compute the

solution. However, we hasten to add that the policy improve-

ment scheme, properly used, will also yield a solution to the

game finitely quick. Namely, let σ select an arbitrary

policy, then τ optimizes, then σ improves his play if

possible, then τ optimizes, etc.; that this scheme termi-

nates can be seen by applying Denardo's [5] analysis to this

case of perfect information. However, we conjecture of this

alternative adjustment scheme that it is only necessary that

both players merely improve, if possible, upon their most

recent play, that is, if σ and τ select arbitrary strategies, then σ improves his play if possible, then τ improves his play if possible, etc., then one still has finite termination. This conjecture is in line with our second question above.

We close this section by treating a model where optimal policies in general lie outside the stationary class. Assume for $\delta \in \Delta$ that $|P_\delta|^n \to 0$ and that the return for using δ and starting in the various states is

$$\left(\sum_{n=0}^{\infty} P_\delta^n\right) R_\delta$$

Note that the return of δ is the fixed point y_δ of L_δ. If $y = L(y) = L_\eta(y)$ then η is Pareto optimal among stationary strategies: that is, there is no δ such that $y \neq y_\delta \geq y$. To see this observe that

$$0 = (P_\delta - I)y_\delta + R_\delta$$

$$0 \leq \max_{\xi \in \Delta} (P_\xi - I)y_\xi + R_\xi$$

$$0 = \max_{\xi \in \Delta} (P_\xi - I)y + R_\xi$$

Letting ξ be the maximizing argument in the second expression we get

$$0 \leq (P_\xi - I)y_\delta + R_\xi$$

$$0 \geq (P_\xi - I)y + R_\xi$$

or

$$(I - P_\xi)y_\delta \leq (I - P_\xi)y$$

Choose $0 \leq \lambda \in R^{1 \times n}$ so that

$$\lambda(I - P_\xi) > 0$$

and we get

$$\lambda(I - P_\xi)y_\delta \leq \lambda(I - P_\xi)y$$

This expression is incompatible with $y \neq y_\delta \geq y$. If Δ is totally regular and if it is also finite, we can compute the unique fixed point.

9. A Barrier Condition

Here we study a sufficient condition for existence of a fixed point of L that is somewhat different from that of §§7 and 8. We then specialize this condition to positive dynamic programming.

Lemma 8: If $L_\delta^z(y) = L^z(y)$ then $L^z(y) + zd_\delta \leq L(y)$

Proof: $L_\delta^z(y) + d_\delta z = L_\delta(y) \leq L(y)$ ☒

Lemma 9: Let $P_{\delta_0} = 0$ and let $(y(t), z(t))$ be the path of the algorithm. If $z(s) > 0$ and $y_i(s + \epsilon) > (R_{\delta_0})_i$ for abritrarily small $\epsilon > 0$ then

$$(L^{z(s)}(y(s)))_i < (L(y))_i$$

Proof: Choose δ so that $y(t) = L_\delta^{z(t)}(y(t)) = L^{z(t)}(y(t))$ for $t = s + \varepsilon$ for small $\varepsilon > 0$. If δ agrees with δ_o at state i we have $y_i(t) = (R_{\delta_o})_i$ for $t = s + \varepsilon$ which contradicts our hypothesis, hence δ does not and we have

$$(L^{z(s)}(y(s)))_i + z(s) \leq (L(y))_i$$

by Lemma 8 ⊠

For $x \in R^n$ let $x^\oplus$ be the set $\{y \in R^n : y \geq x\}$. We say that $T \subset x^\oplus$ is monotone in $x^\oplus$ if $y \in T$, $y' \leq y$, and $y' \in x^\oplus$ imply that $y' \in T$. By the boundary ∂T of T in $x^\oplus$ we mean the set $T \cap$ (closure $x^\oplus \sim T$).

Theorem 10: Suppose that $P_{\delta_o} = 0$, that there is a compact monotone set T in $R_{\delta_o}^\oplus$ with $L(\partial T) \subset T$, and that Δ is finite, then the algorithm yields a fixed point of L in T , that is, $z(t) = 0$ for some $y(t) \in T$.

Proof: We have $y(0) = R_{\delta_o}$ and $R_{\delta_o} = L_{\delta_o}(y(t)) = L_{\delta_o}^{z(t)} \leq y(t)$ for $t \in R^1$. If $z(t) = 0$ for some $y(t) \in T$ then $y(t)$ is a fixed point of L in T . Suppose that $z(t) > 0$ for all $y(t) \in T$; then $y(t)$ leaves

T . Here for some t = s we have y(s) ∈ ∂T and
y(s + ε) ∉ T for small ε > 0 . Let

$$\gamma = \{i : y_i (s + \varepsilon) > y_i(s) \text{ for small } \varepsilon > 0\} .$$

By Lemma 9 we have

$$y(s) = L^{z(s)}(y(s)) \leq L(y(s))$$

with strict inequality at coordinates γ ; hence we have

$$y(s + \varepsilon) = L^{z(s + \varepsilon)}(y(s + \varepsilon) \leq L(y(s))$$

for small ε > 0 . Since L(y(s)) ∈ T we have y(s + ε)
∈ T for small ε > 0 . This contradicts our assumption
that z(t) ≠ 0 for y(t) ∈ T ☒

   If, in the theorem, instead of assuming Δ is finite we
assume that L is finite, then one can easily show that L
has a fixed point by using Brouwer's theorem and the map
εI + (1 − ε)L , also see [9].

   To apply our results to positive dynamic programming as-
sume that $(P_{\delta_o}, R_{\delta_o}) = 0$, that $P_\delta \geq 0$ for δ ∈ Δ , that
Δ is finite, and that for some u ≥ 0 we have L(u) ≤ u ,
see Blackwell [2], Ornstein [17], and Veinott (22). We desire to compute the minimum fixed point y of L .

   Let $T = u^\oplus$ and we have L(T) ⊂ T . We generate the
path (y(t), z(t)) with the algorithm and we claim that
y = y(s) where s is the smallest t such that z(t) = 0 .
Suppose that L(y') = y' ≤ y , let T' = {x : x ≤ y'} , and

we have $L(T') \subset T'$. Then according to Theorem 10 we see that $y \leq y'$ or $y = y'$.

Veinott [22] has shown that this minimum fixed point can be obtained by linear programming and by policy improvement.

10. Comments

There are other areas in dynamic programming as the various time averaging criteria (the first such study is Bellman [1] which might be interesting to pursue with the technique of this paper. One such investigation has led to a decomposition for negative dynamic programs; following decomposition one can apply policy improvement or linear programming to find an optimal policy if it exists, see Strauch [20], Veinott [22] and [11].

The initial motivation for this paper was the notion that complementary pivot theory had something to contribute to dynamic programming (or vice versa). This is a rather natural expectation in view of the facts that linear programming has occupied an important role in dynamic programming and that the complementary pivot algorithms are in some sense more powerful than, say, the simplex method.

Acknowledgements

The author would like to express his indebtedness to R. Grinold, who is jointly responsible for the remarks regarding Shapley's game, to H. Scarf, who helped in showing that solving certain cases of (*) is not a matter of linear programming, and to A. F. Veinott, Jr., who has enhanced the author's thoughts on many aspects of this paper.

This research was supported in part by Army Research Office-Durham Contract DAHC-04-71-C-0041 and NSF Grant GP-34559.

REFERENCES

[1] Bellman, R., "A Markovian Decision Process", Journal of Mathematics and Mechanics, Vol. 6, No. 5, pp. 679-684, 1957.

[2] Blackwell, D., "Positive Dynamic Programming", in Proceedings of Fifth Berkeley Symposium on Mathematical Statistics and Probability, Vol. 1, University of California Press, Berkeley, California, pp. 415-418, 1967.

[3] Blackwell, D., "Discrete Dynamic Programming", Ann. Math. Stat., Vol. 33, No. 2, pp. 719-726, 1962.

[4] Cottle, R. W., and Dantzig, G. B., "A Generalization of the Linear Complementarity Problem", Journal of Combinatorial Theory, Vol. 8, No. 1, January 1970.

[5] Denardo, E. V., "Contraction Mappings in the Theory Underlying Dynamic Programming", SIAM Review, Vol. 9, No. 2, April 1967.

[6] D'Epenoux, F., "A Probabilistic Production and Inventory Problem", Management Science, Vol. 10, No. 1, pp. 98-108, 1963. (Translation of an article published in Revue Francaise de Recherche Operationelle, 14, (1960).)

[7] Derman, C., Finite State Markovian Decision Processes, Academic Press, New York, 1970.

[8] Eaves, B. C., "The Linear Complementarity Problem", Management Science, Vol. 17, No. 9, May 1971.

[9] Eaves, B. C., "On the Basic Theorem of Complementarity", Math. Prog., Vol. 1, No. 1, October 1971.

[10] Eaves, B. C., "Homotopies for Computation of Fixed Points", Math. Prog., Vol. 3, No. 1, August 1972, 1-22.

[11] Eaves, B. C., and Veinott, A. F., Jr., "Policy Improve-
      ment in Stopping Negative, and Positive Markov Decision
      Chains", to appear.

[12] Grinold, R., "A Generalized Discrete Dynamic Program-
      ming Model", CP 340, Center for Research in Management
      Science, School of Business, University of California,
      Berkeley, August 1971.

[13] Howard, R. A., Dynamic Programming and Markov Proces-
      ses, Wiley, New York, 1960.

[14] Kato, T., Perturbation Theory for Linear Operators,
      Springer-Verlag, New York, 1966.

[15] Lemke, C. E., "Recent Results on Complementarity Prob-
      lems", Nonlinear Programming, Academic Press, New York,
      1970.

[16] Lemke, C. E., "Bimatrix Equilibrium Points and Mathe-
      matical Programming", Management Science, Vol. 11,
      No. 7, May 1965, pp. 681-689.

[17] Ornstein, D., "On the Existence of Stationary Optimal
      Strategies", Proc. Amer. Math. Soc., Vol. 20, No. 2,
      pp. 563-569, 1969.

[18] Scarf, H., "An Algorithm for a Class of Nonconvex Pro-
      gramming Problems", Cowles Foundation Discussion Paper
      No. 211, July 14, 1966.

[19] Shapley, L. S., "Stochastic Games", Proc. N.A.S., Vol.
      39, 1953.

[20] Strauch, R. E., "Negative Dynamic Programming", Ann.
      Math. Stat., Vol. 37, No. 4, 871-890, 1966.

[21] Veinott, A. F., Jr., "Discrete Dynamic Programming with
      Sensitive Discount Optimality Criteria", Annals of
      Mathematical Statistics, Vol. 40, No. 5, 1635-1660,
      1969.

[22] Veinott, A. F., Jr., Unpublished notes.

Application of a Fixed Point
Search Algorithm to Nonlinear Boundary
Value Problems Having Several Solutions

E. L. Allgower

ABSTRACT

A fixed point search algorithm is outlined and utilized
to approximate the solutions of finite dimensional analogues
of quasilinear elliptic and ordinary differential equation
boundary value problems having several solutions. Two kinds
of finite dimensional analogues are considered: finite differences and finite orthogonal expansions. In either case
the finite dimensional analogue may be recast as a finite dimensional fixed point problem. If a mapping has several
fixed points, the problem of approximating them by iterative
methods is often severely complicated by such difficulties as
locating appropriate starting points and magnetic properties
of the fixed points. The search type algorithm used here is
generally impervious to all such complications. Several
boundary value problems having a finite number of isolated
solutions are treated numerically and it is found that for

sufficiently fine meshes, the search algorithm applied to the finite difference analogue yields approximations to all of the solutions.   On the other hand, the search algorithm applied to the finite Fourier expansion analogue has not always yielded approximations to all of the solutions of a boundary value problem.

1.   In this paper we treat the problem of obtaining numerical approximations to nonlinear boundary value problems having several isolated solutions by utilizing an algorithm given recently [15] for approximating fixed points of a continuous mapping f of the n-dimensional unit cube C^n into itself. The types of problems which are treated in this way include

$$u'' = f(x,u,u'), \quad u(0) = 0 = u(1); \tag{1.1}$$

$$u'' - u = f(x,u,u'), \quad u(0) = u(1), \quad u'(0) = u'(1);$$

$$u'' + (k/x)u' = f(x,u,u'), \quad u'(0)=0=u(1), \quad -2 < k \text{ (k constant)};$$

$$\nabla u = f(x,u,\nabla u) \text{ on } D, \quad u|\partial D = 0 \text{ on } D ,$$

(D a bounded domain in R^2 with piecewise analytic boundary)

where either f is continuous, and bounded or f is defined on a bounded set arising, for example, from a priori esti- mates on the solutions. It is well known that the boundary value problems (1.1) may be equivalently recast as fixed point problems (i.e., Hammerstein integral equations)

$$Tu(x) = \int_D g(x,y)f(y,u(y),\nabla u(y))dA_y = u(x) \tag{1.2}$$

where g is the Green's function for the differential opera- tor on the left-hand side of (1.1) for the corresponding do- main ([0,1] or D). Existence theorems for solutions to (1.1) in an appropriate Banach space may be established by

application of the Schrauder-Tychonoff theorem. Such existence theorems are not constructive in nature since this
technique offers no means of explicitly constructing the solutions to (1.2) or (1.1). By utilizing a fixed point approximation algorithm such as that given in [15], we may approximate solutions to finite dimensional analogues of (1.2).
We thus obtain sequences of functions which can be shown by
compactness considerations to have subsequences which converge to solutions to (1.2). Basically there are two alternative finite dimensional approaches available leading to finite dimensional analogues of (1.2), viz. discretizations
(e.g., finite differences) and Galerkin methods (e.g., finite
orthogonal expansions). We shall sketch how both approaches
may be made amenable to the fixed point approximation method
discussed in §2, and discuss the computational experience we
have had with both approaches using the numerical procedure
outlined in §2.

It is well known, [9], [12], that if a mapping $f:C^n \to C^n$
has several fixed points, certain of them may be very difficult to approximate by traditional iterative methods. It is
frequently the case that under iterative methods, certain
fixed points of a mapping may be attractive while others may
be repulsive. That is, iteratively defined sequences might
not converge to a repulsive fixed x regardless of which
starting point $x_o \neq x$ is chosen. In such cases the approximation of all of the fixed points of a mapping f by standard methods is often a difficult task. (For examples of
this, see [9], [12].) We shall study certain aspects of the
algorithm we use to conclude that it is essentially impervious to "magnetic" properties of the fixed points of a mapping.
Thus the primary advantage of utilizing the present algorithm

to finite dimensional analogues of (1.2) is that under very general circumstances, all of the fixed points of (1.2) may be approximated. Computational experience bears this out, particularly in the case of finite difference approximations.

2. We now review aspects of the fixed point algorithm in [15] so as to gain insights into how it functions and which fixed points of a mapping may be approximated by the algorithm. The computer program implementing this algorithm is an extension of that given in [1] which in turn is closely related to an algorithm given by Kuhn [18].

A continuous mapping $f = (f_1, \ldots, f_n)$ of C^n into C^n induces a decomposition of C^n into $n + 1$ nonempty closed sets

$$A_o = \text{cl}\{Y \in C^n |\ f_1(Y) \geq Y_1 \neq 1\}, \quad (Y = (Y_1, \ldots, Y_n));$$

$$A_i = \text{cl}\{Y \in C^n |\ f_j(Y) \leq Y_j \neq 0,\ j = 1, \ldots, i \ \text{and} \ f_{i+1}(Y) >$$
$$Y_{i+1} \ \text{or} \ Y_{i+1} = 0\}, \ i = 1, \ldots, n-1;$$

$$A_n = \text{cl}\{Y \in C^n |\ f_j(Y) \leq Y_j \neq 0,\ j = 1, \ldots, n\} \quad (\text{cl} = \text{closure}).$$

We assume the norm on C^n to be $\| Y \| = \max_{1 \leq i \leq n} |Y_i|$. By Sperner's Lemma there exist n-dimensional simplexes σ_n of arbitrarily small diameter such that

$$\sigma_n \cap A_i \neq \phi, \quad i = 0, 1, \ldots, n \quad .$$

Hence by Lebesgue's Lemma, $\bigcap_{i=o}^{n} A_i \neq \phi$. By the manner in which the sets A_i are defined, $Y \varepsilon \bigcap_{i=o}^{n} A_i$ implies $f(Y) = Y$. The sets A_i in fact determine the manner in which the algorithm we use locates a simplex σ_n , which we call a <u>Sperner simplex</u>. For a given positive integer P , the cube C^n is subdivided simplicially into simplexes of diameter $1/P$ as follows. The cube C^n is partitioned into n-dimensional cubes having side length $1/P$ by the hyperplanes $Y_i = j/P$, $i = 1,\ldots,n; j = 1,\ldots,P-1$. Then each subcube $C^n(Y^o,P) = \prod_{i=1}^{n} [Y_i^o, Y_i^o + 1/P]$ is partitioned into n! simplexes of dimension n ,

$$S(i_1,\ldots,i_n;Y^o,P) = \{Y \varepsilon C^n(Y^o,P) | Y_{i_1} \geq Y_{i_2} \geq \ldots \geq Y_{i_n}\}$$

for all possible permutations $(i_1,\ldots,i_n)$ of the integers $(1,\ldots,n)$. Two properties of a simplicial subdivision which play a crucial role in the algorithm for locating a Sperner simplex are: For any n ,

(i) Any (n-1)-dimensional face of a simplex S_n in a simplicial subdivision either lies in ∂C^n or is a common face of exactly two simplexes in the simplicial subdivision.

(ii)   If $S_n \cap A_i \neq \phi$, $i = 0,\dots,n-1$ and $S_n \cap A_n = \phi$,

then S_n has exactly two $(n-1)$-dimensional faces

$S_{n-1}^{(1)}$, $S_{n-1}^{(2)}$ satisfying $S_{n-1}^{(j)} \cap A_i \neq \phi$,

$i = 0,\dots,n-1$; $j = 1,2$.

These two properties uniquely determine the exchange procedure for adjacent simplexes. The algorithm for locating
Sperner simplexes proceeds as follows. The point $Y = 0$ corresponds to the 0-dimensional Sperner simplex $\{0\}$. Now
the 1-dimensional simplex with vertices $0, 0 + Y_P E_1$ (E_i

denotes the ith n-dimensional unit vector). The search com-
mences with the process of exchanging 1-dimensional simplexes via 0-dimensional faces (which are 0-dimensional
Sperner simplexes) until a 1-dimensional Sperner simplex
with vertices $((j-1)/P)E_1$, $(j/P)E_1$ is found (i.e.,

$S_1 \cap A_i \neq \phi$, $i = 0,1$) . Now the 2-dimensional Sperner sim-

plex with vertices $Y^O = (j-1)/PE_1$, $Y^O + 1/P \sum_{i=1}^{j} E_i$, $j = 1,2$

is formed and used as the starting point for the exchange process until a 2-dimensional Sperner simplex is located. This
process is repeated until an n-dimensional Sperner simplex
has been located. It may be shown [2], [18] that although it
is necessary to reduce the dimension in the above process if
$S_{n-1}^{(2)} \subset \partial C^n$, the procedure must terminate in a finite number

of exchanges by locating an n-dimensional Sperner simplex.
From the preceding we see that the algorithm describes a
marching process in steps of size $1/P$ along sections of the
sets

$$\bigcap_{i=o}^{k} A_i, \qquad k = 0,1,\ldots,n \qquad\qquad (2.1)$$

successively until a Sperner simplex is reached.

Once a Sperner simplex for this simplicial subdivision has been located, we implement the preceding process to seek additional Sperner simplexes by permuting the sets A_i cyclically, setting $A_i \rightarrow A_{i+j \pmod{n+1}}$, $j = 1,\ldots,n$ and exchanging via the $(n-1)$-face S_{n-1} satisfying $S_{n-1} \cap A_i \neq \phi$, $i = 0,1,\ldots,n-1$ after the permutation has been performed. Now the procedure either locates another Sperner simplex or must eventually reduce dimensions twice. In either case, we then return to the previous Sperner simplex, perform another permutation (if possible) and proceed again until all possible exits from all located Sperner simplexes have been performed. Often it is found to be advantageous to perform a less exhaustive search such as permitting only one exit from each Sperner simplex or stopping after an expected number of Sperner simplexes has been located. The reason for not permitting a reduction of dimension more than twice in the search for subsequent Sperner simplexes is to prevent a cycle from occurring and also to avoid excessive inefficiency. Figure 2.1 portrays the above procedure for a special case with $n = 2$.

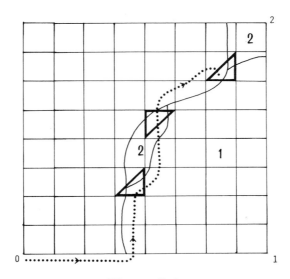

Figure 2.1

The following theorem indicates how nearly fixed the points in a Sperner simplex must be.

<u>Theorem 2.1.</u> [15] Suppose that $f:C^n \rightarrow C^n$ and $\| f(Y) -$ $f(X) \| \leq \varepsilon$ whenever $\| Y - X \| \leq \delta$. Let S^n be an n-dimensional Sperner simplex whose diameter is $\leq \delta$. Then $\| f(X) - X \| \leq \varepsilon + \delta$ for any $X \varepsilon S^n$.

The next theorem gives a sufficient condition for a fixed point Y of f to have a sequence $S^n(P)$ of diameter Y_P converge to Y .

<u>Theorem 2.2.</u> [5] Let $B_r(Y) = \{X \varepsilon C^n | \| X - Y \| < r\}$ and $Y \varepsilon \overset{n}{\underset{i=0}{\cap}} A_i$ be such that each of the sets $B_r(Y) \cap A_i$, i = 0,...,n contains an open set. Then for a sufficiently

large integer P_0 , there exists a sequence of Sperner sim-

plexes $\{S_n(P)\}_{P \geq P_0}$ such that $S_n(P) \to Y$ as $P \to \infty$.

Remark 2.1. There may be a fixed point $Y \in \bigcap_{i=0}^{n} A_i$ of f

satisfying the hypothesis of Theorem 2.2 for which the
Sperner simplexes converging to Y will never be attained by
the above search procedure. In fact, we attain only those
Sperner simplexes which may be reached by traversing a particular path which lies in ∂C^n and the sets (2.1), and has 0
as the initial point.

Remark 2.2. On the other hand, even if f is not a self-map
of C^n into C^n , any Sperner simplex which may be reached
by such a particular path will be attained by the search procedure. This fact is used below to develop an iterative extension of the search algorithm, and to approximate certain
fixed points whose existence is not implied by the Brouwer
fixed point theorem.

Remark 2.3. If f admits a linearization close to a fixed

point $Y \in \bigcap_{i=0}^{n} A_i$, then the sequence $\{S_n(P)\}_{P \geq P_0}$ in

Theorem 2.2 converges linearly to Y as $P \to \infty$ [21].

We now sketch briefly how the preceding algorithm may
be restarted to obtain refined Sperner simplex approximations

to a particular attainable point in $\bigcap_{i=0}^{n} A_i$. Our device is

related to the devices due to Merrill [19] and Eaves [13].

Let $P \geq 2$ be a fixed positive integer and define

$$C^n(j) = \{(Y_1,\ldots,Y_n,j) \mid 0 \leq Y_k \leq P^j, \ k = 1,\ldots,n\}$$

$$B^n(j) = \{(Y_1,\ldots,Y_n,j) \mid 0 \leq Y_k \leq P^{j+1}, \ k = 1,\ldots,n\}, \ j = 0,1,\ldots \ . \tag{2.2}$$

We decompose each set $B^n(j)$ into closed sets $A_s(j)$, $s = 1,\ldots,n$ so that the sets $A_s(j) \cap C^n(j)$ are linearly homeomorphic by a factor P^j to A_s, $s = 0,1,\ldots,n$, respectively. On $B^n(j) - C^n(j)$, we assign

$$Y \in A_0(j) \quad \text{if} \quad Y_1 = 0$$

$$Y \in A_i(j) \quad \text{if} \quad Y_k > P^j \ \text{for some} \ k = 1,\ldots,n, \ Y_k > 0 \ \text{for} \tag{2.3}$$

$$k = 1,\ldots,i \quad \text{and} \quad Y_{i+1} = 0 \ .$$

Now the original search algorithm is used (with $P = 1$) to seek a Sperner simplex of dimension $n + 1$ in the positive orthant of R^{n+1} . (Here is where we utilize the fact observed in Remark 2.2.) There is obviously no such simplex since $A_{n+1}(j) = \phi$ for $j \geq 0$. Owing to (2.2) and (2.3), the search is forced to proceed into higher hyperplanes $Y_{n+1} = j$. At each stage where an exchange to a higher hyperplane $Y_{n+1} = j + 1$ is performed, the exchange is made via a face which is an n-dimensional Sperner simplex $S^n(j)$ in $C^n(j)$. Under the inverse linear homeomorphism, $S^n(j)$ corresponds to a Sperner simplex in C^n of diameter $(1/P)^j$.

3.   In this section we carry out the finite difference approach to solving (1.1) by means of the fixed point algorithm described in §2.   We shall indicate the implementation of the approach and discuss certain numerical experience.   In the interest of brevity we omit the convergence proofs.   They are carried out in detail in [3] and [4].   We shall formulate the treatment here so as to include all of the cases in (1.1), explicit details for the individual cases in (1.1) are to be found in [3], [4].

We assume that a uniform (square) grid of mesh size   h has been placed upon   [0,1]   or   D   and that all derivatives are approximated by central differences.   The grid points in D   are ordered in the "natural order", from left to right and top to bottom: $P_1, P_2, \ldots, P_n$.   Assuming the boundary conditions to have been incorporated, the finite difference analogue of (1.1) may be written

$$A_h \bar{u} = h^2 \bar{f}(\bar{u}) \tag{3.1}$$

where $\bar{u} = (u(P_1), \ldots, u(P_n))'$, $\bar{f}(\bar{u}) = (f(P_1, u(P_1), \nabla_h u(P_1))$ $,\ldots, f(P_n, u(P_n), \nabla_h u(P_n)))'$, A_h is a monotone matrix [20] arising from the discretization of the differential operator and the boundary conditions.   In particular, $A_h^{-1} = (a_{ij}^{-1})$ exists and $a_{ij}^{-1} \geq 0$, $i,j = 1, \ldots, n$.   Hence the solutions to (3.1) are the solutions to

$$\bar{u} = h^2 A_h^{-1} \bar{f}(\bar{u}) \tag{3.2}$$

and conversely. Thus the solutions to (3.1) are the fixed

points of the continuous mapping $T_n : R^n \to R^n$,

$$T_n \bar{u} = h^2 A_h^{-1} \bar{f}(\bar{u}) \qquad (3.3)$$

and conversely. If f is bounded, say $|f| \leq M$ on its do-

main (e.g., $|f| \leq M$ on $\bar{D} \times R^3$) , then

$$|T_n \bar{u}(\bar{P})| \leq h^2 M A_h^{-1} \bar{e}$$

where $\bar{e}$ is the n-dimensional vector whose components are

all unity. Hence an a priori bound on any fixed point $\bar{u}$ of

T_n is given by

$$|u_i| = |u(P_i)| \leq h^2 M \sum_{j=1}^{n} a_{ij}^{-1} , \quad i = 1,\ldots,n . \qquad (3.4)$$

If we let $B^n = \{\bar{v} \in R^n | |\bar{v}| \leq h^2 M A^{-1} \bar{e}\}$, then $T_n : R^n \to B^n$.

In particular T_n is a continuous mapping of B^n into it-

self. Hence a simple linear homeomorphism $H : B^n \to C^n$ yields

a mapping

$$G = H T_n H^{-1} \qquad \text{of} \qquad C^n \quad \text{into} \quad C^n .$$

That is, if $a_i \leq u_i \leq b_i$, $i = 1,\ldots,n$ and $T_n : \prod_{i=1}^{n} [a_i, b_i] \to$

$$\prod_{i=1}^{n} [a_i, b_i] \quad , \text{ then set } \quad Y_i = \alpha_i u_i + \beta_i \quad \text{with} \quad \alpha_i = 1/(b_i - a_i),$$

$\beta_i = -a_i \alpha_i$ and

$$g_i(Y_1, \ldots, Y_n) = \alpha_i T_n^i \left(\frac{Y_1 - \beta_1}{\alpha_1}, \ldots, \frac{Y_n - \beta_n}{\alpha_n} \right), \quad i = 1, \ldots, n \quad .$$

For the purpose of numerical computations, we do not use the mapping T_n , but rather the mapping

$$T_n' \bar{u} = (I - D_A^{-1} A_h) \, \bar{u} + h^2 D_A^{-1} \bar{f}(\bar{u}) \tag{3.5}$$

where $D_A = \text{diag } A_h$ and I is the $n \times n$ identity matrix. The reason for this choice is simply that, in general, A_h is a sparse matrix while on the other hand, A_h^{-1} is generally a full matrix. Thus the arithmetical evaluations may be more efficiently performed for T_n' than for T_n . The mapping T_n' is also a continuous self-mapping of B^n since for $\bar{u} \, \varepsilon \, B^n$,

$$|T_n' \bar{u}| \leq |(I - D_A^{-1} A_h) \bar{u}| + h^2 |D_A^{-1} \bar{f}(\bar{u})| \leq h^2 M(I - D_A^{-1} A_h) A_h^{-1} \bar{e} +$$

$$h^2 M D_A^{-1} \bar{e} = h M A_h^{-1} \bar{e} \quad .$$

In establishing this we have used the fact that A_h is diagonally dominant.

In contrast with the method which we discuss in the next section, the approach used here does not require any explicit formulation of the Green's function g in (1.2). Recently it has been shown [17] that under very general circumstances (which have always been fulfilled in our examples), if the ordinary differential equations in (1.1) have a finite number of isolated solutions, then for each isolated solution of (1.1), the difference scheme has, for a sufficiently fine mesh a unique solution in some tube about the isolated solution. It is also shown that the numerical solution can be computed by Newton's method with quadratic convergence. However, the usual difficulty of obtaining appropriate starting values is still present. Computational experience with the present algorithm also bears out the result in [17]. That is, in our examples as many solutions to (1.1) as are known to be present have been approximated. The approximations obtained by the present algorithm might be used as starting vectors for Newton's method if particularly sharp results are desired.

Example 3.1.   The generalized Duffing equation for the motion of a pendulum is

$$u'' + \lambda^2 \sin u = 0 \quad , \quad u(0) = 0 = u(\pi) \quad .$$

In this case, $u \equiv 0$ is a solution and if u is a solution, then so is $-u$. If $k - 1 < \lambda < k$ (k a positive integer), there are $2k - 1$ solutions. For $n = 2$, the mapping T'_n induces a decomposition of C^n such as that shown in Figure 2.1. As λ was allowed to increase through moderate increments, the expected number of solutions was

always obtained. This suggests that the present algorithm
may also be used to locate bifurcation points. It is well
known that bifurcation may only take place at points in the
spectrum of an operator, but that bifurcation need not necessarily take place. In this example certain instabilities
occurred near bifurcation points such as temporary loss of
symmetry and occasionally several simplexes approximated the
same fixed point.

Example 3.2. Two nonlinear boundary value problems arising
in the theory of gas dynamics and chemical reactors [14],
[16] are

$$u'' + \lambda e^u = 0, \ u(0) = 0 = u(1) \ \text{(Bratú equation)},$$

$$u'' + (k/x)u' + \lambda e^u = 0, \ u'(0) = 0 = u(1) \ \text{(Emden-type equation)} \ .$$

In each of these cases, for $0 \le k \le 2$, there is a critical
value λ_c such that for $0 < \lambda < \lambda_c$, there will be two
solutions $0 < u_1 < u_2$ on $(0,1)$. As $\lambda \to \lambda_c$ the two
solutions converge to a single solution u_c . For $\lambda > \lambda_c$
there are no solutions. Demonstrations of these results may
be found in [14] and [16]. An interesting feature of this
example is that under many iterative procedures, the upper
solution corresponds to a repulsive fixed point of T and
this property also carries over to T_n and T'_n . By using
the algorithm described in §2, both solutions are approximated when the mesh on C^n (i.e., 1/P) is taken sufficiently
small. We note that for any n-cell B^n which contains both

solutions, T_n and T_n' are not self-mappings on B^n .

When $\lambda = \lambda_c$, the discrete mapping T_n' will yield a decomposition in which A_n will be a single point and hence the present algorithm will in general not succeed in locating u_c . On the other hand, the critical value λ_c may be approximated by incrementing λ so that λ approaches λ_c , i.e., $u_1 \uparrow u_c$ and $u_2 \downarrow u_c$. The point λ_c is sometimes referred to as a secondary bifurcation point.

For a particular value of λ in the Bratú problem, the lower solution was approximated for $n = 25$, $P = 128$ in less than 4 seconds, while for a search with restart the same simplex was attained in less than 2 seconds using a CDC 6600 computer at the Aerospace Research Laboratories at Wright-Patterson AFB.

Example 3.3. An equation which arises in chemical reactor theory [7] is

$$u'' + \alpha(\beta - u)e^{-\gamma/u} = 0, \qquad u'(0) = 0, \qquad u(1) = 1 \ .$$

It has been proven by degree theoretical considerations [6], that for appropriate values of the constants α, β, γ , there will be three solutions $0 < u_1 < u_2 < u_3$ on $(0,1)$. In this event the intermediate solution u_2 corresponds under many iterative algorithms to a repulsive fixed point. The solution u_2 also presents difficulties for Newton-type methods for solving the corresponding system of equations.

The present algorithm succeeds in approximating all three solutions. Numerical data for a particular example appears in [4].

Example 3.4. A 2-dimensional analogue of Example 3.1 is

$$\Delta u + \lambda^2 \sin u = 0 \quad \text{on} \quad D = (-1,1) \times (-1,1),$$

$$u|\partial D = 0 \quad .$$

The eigenvalues in this problem are $\lambda_k = (2k + 1)\pi/2$ and for $\lambda_{k-1} < \lambda < \lambda_k$, there will be $2k + 1$ solutions which occur in symmetric pairs about $u \equiv 0$. In this example the present algorithm was successfully applied to approximate the expected number of solutions for various values of λ .

Example 3.5. A 2-dimensional analogue of Example 3.2 is

$$\Delta u + \lambda e^u = 0 \quad \text{on} \quad D = (-1,1) \times (-1,1),$$

$$u|\partial D = 0 \quad .$$

This nonlinear boundary value problem has analogous applications to those in Example 3.2 and the same qualitative behavior of the solutions obtained viz. there is a critical λ_c such that for $\lambda < \lambda_c$, two solutions with $0 < u_1 < u_2$ on D exist and $u_1 \uparrow u_c, u_2 \downarrow u_c$ as $\lambda \to \lambda_c$. In this example we seek to approximate the solutions u which are symmetric relative to the axes of symmetry of D . Thus we consider the mixed boundary value problem with the boundary condition

$$u = 0 \quad \text{on} \quad \{(X,Y) \mid X = 1, \ 0 \le Y \le 1\} \quad \text{and}$$

$$\frac{\partial u}{\partial n} = 0 \quad \text{on} \quad \{(X,Y) \mid 0 \le X \le 1 \ \text{and} \ Y = 0 \ \text{or} \ Y = X\}$$

where $\partial/\partial n$ denotes the exterior normal derivative at the boundary.

Although the nonlinear part e^u is not bounded, we may obtain a priori bounds on the solutions by using recent results obtained by Bandle [8]. Using such estimates to obtain an n-cell B^n , the algorithm in §2 functions successfully to yield approximations to both of the solutions u_1, u_2 despite the fact that T_n' is not a self-mapping on B^n .

Additional examples which have been treated by this approach are to be found in [3], [4], [15].

4. In this section we present an application of the truncated orthogonal expansion approach to the nonlinear periodic boundary value problem

$$X'' - X = F(t,X,X')$$

$$\tag{4.1}$$

$$X(0) = X(1), \quad X'(0) = X'(1)$$

where $F(t,X,X')$ is a continuous and bounded vector function from $[0,1] \times R^n \times R^n$ into R^n . We outline here the

essential aspects of this approach. The full details are
given by Chen [10], [11].

For the scalar problem

$$X'' - X = 0$$

$$X(0) = X(1), \quad X'(0) = X'(1),$$

the Green's function is

$$g(t,s) = \begin{cases} [(e^{-1} - 1)e^{t-s+1} + (1 - e)e^{s-t-1}]/\Delta, & t \leq s \\ \\ [(e^{-1} - 1)e^{t-s} + (1 - e)e^{s-t}]/\Delta, & s \leq t \end{cases} \quad (4.2)$$

where $\Delta = 2(e + e^{-1} - 2)$. The existence of a fixed point
of

$$TX = \int_0^1 g(t,s)F(s,X,X')ds \qquad (4.3)$$

is equivalent to the existence of a solution to (4.1). The
existence of fixed points of TX , i.e., solutions to (4.1)
in the L^2 sense may be seen from the following discussion.

Let $L^2[0,1]$ denote the real Hilbert space of Lebesgue
square-integrable functions from $[0,1]$ into R^n with the
inner product

$$(Y,Z) = \sum_{k=1}^{n} \int_0^1 Y^{[k]}(s)Z^{[k]}(s)ds \quad \text{and norm} \quad \| Y \|_2 = (Y,Y)^{1/2}$$

where $Y^{[k]}$ denotes the kth component of the vector Y .

Let $\{\Phi_i(t)\}_{i=o}^{\infty}$ be a complete orthonormal sequence of func-

tions in $L^2[0,1]$. An example of such a sequence is the
sequence

$$\Phi_{kn+j}(t) = \begin{cases} (\sqrt{2} \cos k\pi t)E_{j+1}, \ j = 0,\ldots,n-1 \text{ for } k \text{ even} \\ \\ (\sqrt{2} \sin (k+1)\pi t)E_{j+1}, \ j = 0,\ldots,n-1 \text{ for } k \text{ odd} \end{cases} \qquad (4.4)$$

where E_i denotes the ith n-dimensional unit vector.

Let m be a non-negative integer. Let $(A_m^{[o]},\ldots,$
$A_m^{[m]})$ and $(B_m^{[o]},\ldots,B_m^{[m]})$ be arbitrary ordered $(m + 1)$-
tuples of real numbers. It follows from (4.2), the bounded-
ness of F and the Brouwer fixed point theorem that there
exist solution vectors to the systems

$$A_m^{[k]} = (\Phi_k(t), \int_o^1 g(t,s)F(s, \sum_{k=o}^{m} A_m^{[k]}\Phi_k(s), \sum_{k=o}^{m} B_m^{[k]}\Phi_k(s))ds),$$

$$(4.5)$$

$$B_m^{[k]} = (\Phi_k(t), \int_o^1 \frac{\partial g}{\partial t}(t,s)F(s, \sum_{k=o}^{m} A_m^{[k]}\Phi_k(s), \sum_{k=o}^{m} B_m^{[k]}\Phi_k(s))ds)$$

for $k = 0,1,\ldots,m$. Furthermore, solutions to (4.5) may be
approximated by means of the algorithm discussed in §2 when
used in conjunction with a numerical quadrature process.

We now define

$$X_m(t) = \sum_{k=0}^{m} A_m^{[k]} \Phi_k(t) \quad ,$$

$$Y_m(t) = \sum_{k=0}^{m} B_m^{[k]} \Phi_k(t) \quad .$$

It may be shown [10] by means of compactness considerations and Bessel's inequality that there exist subsequences $\{X_{m_k}\}$, $\{Y_{m_k}\}$ and a solution X to (4.1) such that

$$\| X_{m_k}(t) - X(t) \| \to 0 \quad \text{and} \quad \| Y_{m_k}(t) - X'(t) \| \to 0 \text{ as } m_k \to \infty .$$

Moreover, we have

Theorem 4.1.   [10]   Let $S = \{X(t) \mid X(t) \text{ is a solution to } (4.1)\}$, $S' = \{X'(t) \mid X(t) \ \varepsilon \ S\}$.   Then

$$d_2(X_m(t), S) = \inf_{X(t) \varepsilon S} \| X_m(t) - X(t) \| \to 0$$

$$d_2(Y_m(t), S') = \inf_{X'(t) \varepsilon S'} \| Y_m(t) - X'(t) \| \to 0 \quad \text{as} \quad m \to \infty .$$

An advantage which the approach in this section offers is that owing to knowledge of the asymptotic behavior of the coefficients $A_m^{[k]}$, $B_m^{[k]}$ as $m \to \infty$, we may expect relatively good accuracy of the approximations for relatively low

values of m . This has also been evidenced in our computational experience. The orthogonal sequence (4.4) has been
used in all of our numerical experiments. A disadvantage of
this approach is that since numerical quadratures are involved, the functional evaluations of the mappings defined by
(4.5) require more computing time than do our mappings in the
finite difference approach.

    A surprising occurrence has been that this approach has
not always been successful in locating approximations to all
of the solutions known to be present. For example, in Example 3.1, with $\lambda = 2.5$, there are five solutions; the zero
solution u_o , a positive solution u_1 , a solution u_2
which is positive on $(0,\pi/2)$ and negative on $(\pi/2,\pi)$ and
$-u_1,-u_2$. When the algorithm in §2 was used for the mapping
defined by (4.5), the only solutions which were approximated
were $u_1,-u_2,u_2$ and the approximations were obtained in that
order. In the finite difference approach, after the linear
mapping H of B^n onto C^n has been applied, the solutions
written in increasing order of distance from 0 , are
$\bar{u}_1,\bar{u}_2,\bar{u}_0,\bar{u}_1$ ($\bar{u}_i$ denotes the approximation to u_i). This is
also the order in which the solutions are generally obtained
in the finite difference approach. This suggests that the
searching path for all of the solutions of (1.1) is usually
"minimal" for the finite difference approach, while for the
Galerkin approach it is often not minimal. Numerical examples for this approach are presented in [10].

REFERENCES

[1]   Allgower, E. L., and Keller, C. L., "A Search Routine
      for a Sperner Simplex," Computing 8, 157-165, 1971.

[2]   Allgower, E. L., Keller, C. L., and Reeves, T. E., "A
      Program for the Numerical Approximation of a Fixed
      Point of an Arbitrary Continuous Mapping of the N-Cube
      or N-Simplex Into Itself," Aerospace Research Labora-
      tories Report 71-0257, Wright-Patterson Air Force Base,
      Ohio, 1971.

[3]   Allgower, E. L., and Jeppson, M., "The Approximation of
      Solutions of Nonlinear Elliptic Boundary Value Problems
      Having Several Solutions", Numerische, Insbesondere
      Approximationstheoretische Behandlung Von Funktional-
      gleichungen, Lecture Notes in Mathematics 333, 1-20,
      1973.

[4]   Allgower, E. L., and Jeppson, M., "Numerical Solutions
      of Nonlinear Boundary Value Problems", to appear in
      Numer. Math.

[5]   Allgower, E. L., "Numerische Approximation von Lösungen
      Nicht-Linearer Randwertaugaben Mit Mehreren Lösungen",
      ZAMM, 54, T206,207 (1974).

[6]   Amann, H., "Existence of Multiple Solutions for Non-
      linear Elliptic Boundary Value Problems", Indiana Univ-
      ersity Math. J. 21, 925-935, 1972.

[7]   Aris, R., "On Stability Criteria of Chemical Reaction
      Engineering", Chem. Eng. Sci. 24, 149-169, 1969.

[8]   Bandle, C., "Sur un Problem de Dirichlet Non Linéaire,"
      C. R. Acad. Sci. Paris Sér. A-B 276, A1155-A1157, 1973.

[9]   Brown, K. M., and Gearhart, W. B., "Deflation Tech-
      niques for the Calculation of Further Solutions of a
      Nonlinear System", Numer. Math. 16, 334-342, 1971.

[10] Chen, H., "A Constructive Existence Method for Non-linear Boundary Value Problems", submitted to *J. Diff. Eqs.*

[11] Chen, H., "A Constructive Existence Method for Non-linear Boundary Value Problems", Ph.D. Thesis, Colorado State University, 1974.

[12] Collatz, L., "Functional Analysis and Numerical Mathematics", New York, 1967.

[13] Eaves, B. C., "Homotopies for Computation of Fixed Points," *Mathematical Programming* 3, 1-22, 1972.

[14] Gelfand, I. M., "Some Problems in the Theory of Quasi-linear Equations", *Amer. Math. Soc. Transl.* (2) 29, 295-381, 1963.

[15] Jeppson, M., "A Search for the Fixed Points of a Continuous Mapping", *Mathematical Topics in Economic Theory and Computation,SIAM*, Philadelphia, 1972.

[16] Joseph, D. D., and Lundgren, T. S., "Quasilinear Dirichlet Problems Driven by Positive Sources", *Arch. Rational Mech. Anal.* 49, 241-269, 1973.

[17] Keller, H. B., "Accurate Difference Methods for Non-linear Two Point Boundary Value Problems", *SIAM J. Num. Anal.* 11, 305-320, 1974.

[18] Kuhn, H. W., "Approximate Search for Fixed Points", *Computing in Optimization Problems* 2, 199-211, New York, 1969.

[19] Merrill, O. H., "Applications and Extensions of an Algorithm that Computer Fixed Points of Certain Non-empty Convex Upper Semi-continuous Point to Set Mappings", Tech. Rep. 71-7, Dept. of Industrial Engineering, University of Michigan, Ann Arbor, 1971.

[20] Varga, R. S., "Matrix Iterative Analysis", Englewood Cliffs, 1962.

[21] Vertgeim, B. A., "On an Approximate Determination of the Fixed Points of Continuous Mappings", *Soviet Math. Dokl.* 11, 295-298, 1970.

Generating Stationary Points for a Class of
Mathematical Programming Problems by
Fixed Point Algorithms

Terje Hansen

ABSTRACT

The paper presents a fixed point algorithm that generates stationary points for the program

$$\text{maximize} \quad f(x) \quad,$$
$$\text{subject to} \quad g_i(x) \leq 0 \ , \ i = 1,\ldots,m,$$
$$x \geq 0 \quad,$$

where each $g_i(x)$ is a convex function of x .

1. Introduction

The purpose of this paper is to present an algorithm
that generates stationary points for the program

$$\text{maximize} \quad f(x_1,\ldots,x_n),$$

$$g_i(x_1,\ldots,x_n) \leq 0, \quad i=1,\ldots,m, \tag{1}$$

subject to

$$x \geq 0,$$

where each $g_i(x)$ is a convex function of x . Our interest in stationary points for (1) is motivated by the fact
that the set of stationary points contains the optimal solutions to (1) if they exist.

Hansen [2] has suggested a fixed point method for solving (1) in the case where $f(x)$ is concave. Similar approaches may also be found in Eaves [1] and Merrill [4]. The
algorithm to be presented below is essentially an extension
of the one suggested by Hansen [2]. The algorithm is based
on a combinatorial theorem due to Scarf [5].

2. A Combinatorial Theorem due to Scarf

The combinatorial theorem is expressed in terms of a
primitive set of vectors. Let $\Pi = \left\{\pi^1,\ldots,\pi^h\right\}$ be a collection of vectors in n-dimensional Euclidian space of the fol-
lowing kind.

The first n vectors of Π , the so-called slack vectors, have the form:

$$\pi^1 = (0, M_1, M_1, \ldots, M_1)' \quad ,$$

$$\pi^2 = (M_2, 0, M_2, \ldots, M_2)' \quad ,$$

.

.

.

$$\pi^n = (M_n, M_n, M_n, \ldots, 0)' \quad ,$$

where $M_1 > M_2 > \ldots M_n > 1$ and $'$ denotes transposition.
The remaining vectors lie on the unit simplex, i.e., for all
$j > n$, we have,

$$\sum_{i=1}^{n} \pi_i^j = 1, \ \pi_i^j \geq 0, \ i=1, \ldots, n \quad .$$

Finally we make the non-degeneracy assumption that no two
distinct vectors in Π have the same i^{th} coordinate for
any i .

<u>Definition of a Primitive Set</u>: A set $\left\{ \pi^{j_1}, \ldots, \pi^{j_n} \right\}$ of n
distinct vectors in Π is defined to be a primitive set if
there is no vector π^j in Π for which

$$\pi_i^j > \min \left\{ \pi_i^{j_1}, \ldots, \pi_i^{j_n} \right\} \quad \text{for all} \ i \quad .$$

If a specific vector π^{j_α} in a primitive set
$\left\{ \pi^{j_1}, \ldots, \pi^{j_n} \right\}$ is removed there will in general be a unique
replacement π^j in Π so that the new collection of vectors
also forms a primitive set. The non-degeneracy assumption on
the vectors in Π permits the statement of the following
theorem (see Scarf [5] for details of the proof).

Theorem: Let the set $\left\{\pi^{j_1}, \ldots, \pi^{j_n}\right\}$ of distinct vectors in Π be a primitive set. Then aside from one exceptional case there is a unique vector π^j in Π which yields a primitive set when it replaces π^{j_α}. The exceptional case arises when the primitive set consists of $n-1$ vectors from the first n vectors of Π and one vector π^j with $j > n$ and we are attempting to remove the latter vector. In this case no replacement is possible.

The combinatorial theorem is expressed in terms of the concept of a primitive set.

Combinatorial Theorem: Let A be a matrix with n rows and h columns with the first n columns forming an identity matrix and let the jth column of A be associated with the jth member of Π as follows:

$$
\begin{array}{cccc}
\pi^1 & \pi^2 \ldots \ldots \pi^n & \pi^{n+1} \ldots \ldots \pi^h &
\end{array}
$$

$$
A = \begin{bmatrix}
1 & 0 \ldots \ldots 0 & a_{1,n+1} \cdots \cdots a_{1,h} \\
0 & 1 \ldots \ldots 0 & a_{2,n+1} \cdots \cdots a_{2,h} \\
\cdot & \cdot \quad\quad \cdot & \cdot \quad\quad\quad \cdot \\
\cdot & \cdot \quad\quad \cdot & \cdot \quad\quad\quad \cdot \\
\cdot & \cdot \quad\quad \cdot & \cdot \quad\quad\quad \cdot \\
0 & 0 \ldots \ldots 1 & a_{n,n+1} \cdots \cdots a_{n,h}
\end{bmatrix}
$$

Let $b = (b_1, \ldots, b_n)'$ be a non-negative vector such that the set of non-negative solutions to the system of equations $Ay = b$ is bounded. Then there exists a primitive set

$\left\{\pi^{j_1}, \ldots, \pi^{j_n}\right\}$ so that the corresponding columns $a_{j_1}, \ldots,$
a_{j_n} of A form a feasible basis (in the linear programming sense) for $Ay = b$.

3. Proof of the Combinatorial Theorem

We shall find it useful to review Scarf's proof of the combinatorial theorem. The theorem is proved by an algorithm. The algorithm alternates between the operation of replacing a vector in a primitive set and a pivot step on the system of equations $Ay = b$.

Since by assumption the set $\{y \mid y \geq 0, Ay = b\}$ is bounded, an arbitrary column outside of a feasible basis can be brought into the basis by a pivot step, and if we make the standard non-degeneracy assumption of linear programming, a unique column will be removed.

Let us begin with a primitive set consisting of the vectors $\pi^2, \ldots, \pi^n$ and a single vector π^j with $j > n$. In order for this to yield a primitive set π^j must be that vector in Π (other than the first n members of Π), with the largest first coordinate. On the other hand the columns $1, 2, \ldots, n$ form a feasible basis for $Ay = b$, since the bector b is assumed to be non-negative.

At each iteration the algorithm will typically be in a position similar to the following: The primitive set will contain the vectors $\pi^{j_1}, \ldots, \pi^{j_n}$ with none of them equal to π^1 , and the feasible basis for $Ay = b$ will be given by

the columns $(1, j_2, \ldots, j_n)$. The two sets of indices, which
we wish to make identical in all coordinates, will in fact be
equal in n-1 coordinates, and differ in the remaining one.
At each position other than the first one there will be two
operations which lead to a similar state.

One possibility is to remove the vector π^{j_1} from the
primitive set, and either to terminate if π^1 is introduced
into the primitive set, or to be lead to a new position in
which n-1 of the coordinates are identical. The other possibility is to introduce column j_1 into the feasible basis
for Ay = b , and either to terminate if column 1 is removed, or again to be lead to a new position with n-1 identical coordinates. In the original position only one of
these two operations can be carried out since the other alternative is the exceptional case mentioned in Section 2. In
any subsequent position we select that alternative other than
the one used in arriving at that position.

The Lemke-Howson argument [3] demonstrates that the
algorithm cannot cycle, since if the first position to be
repeated occurs in the middle of the algorithm there would
necessarily be three alternatives available at that step
rather than two. And if the first position to be repeated is
the original one, there would necessarily be two alternatives,
rather than one. Since the number of positions is finite,
the algorithm must terminate, and it can only do so with a
solution to the problem. This concludes the proof of the
combinatorial theorem.

4. <u>Construction of an Algorithm to Approximate a Stationary</u>
    <u>Point</u>

Consider the problem

maximize $f(x_1,\ldots,x_n)$,

subject to $g_i(x_1,\ldots,x_n) \leq 0$, $i=1,\ldots,m$,

$x \geq 0$.

We shall make the following stipulations on the problem:

i)   The functions $g_i(x)$, $i=1,\ldots,m$, are convex.

ii) There is a constant $0 < c < 1$ so that if
     $x = (x_1,\ldots,x_n)$ is a non-negative vector
     satisfying the constraints $g_i(x) \leq 0$ for
     $i=1,\ldots,m$ then $x_1+\cdots+x_n \leq c$. This as-
     sumption only requires that the constraint
     set be bounded.

iii) There exists a non-negative vector x^* , with
     $g_i(x^*) < 0$ for all i . This is the customary
     constraint qualification.

iv) The functions $f(x)$, $g_i(x)$, $i=1,\ldots,m$,
     have continuous partial derivatives in the set
     $\left\{ x \mid x_i \geq 0, \sum_1^n x_i \leq 1 \right\}$. We shall represent
     the partial derivative of $f(x)$ and $g_i(x)$
     with respect to x_j by $f_j(x)$ and $g_{ij}(x)$
     respectively.

The non-negative vector $\bar{x} = (\bar{x}_1, \ldots, \bar{x}_n)$ is a stationary point for (1) if there exists a non-negative vector $(\bar{\lambda}_1, \ldots, \bar{\lambda}_m)$ such that $\bar{x}$ and $\bar{\lambda}$ satisfy the inequalities:

$$f_j(\bar{x}) - \sum_{i=1}^{m} \bar{\lambda}_i g_{ij}(\bar{x}) \leq 0 \quad , \text{ for all } j \quad ,$$

(2.1)

with equality if $\bar{x}_j > 0$, and

$$g_i(\bar{x}) \leq 0 \quad , \text{ for all } i \quad ,$$

(2.2)

with equality if $\bar{\lambda}_i > 0$.

In order to apply the combinatorial theorem presented in the preceding section to approximate a stationary point for (1), we shall introduce an additional coordinate $x_o = 1 - \sum_{i=1}^{n} x_i$ and work with the simplex of dimensionality $n+1$,

$$S = \left\{ x \mid x \geq 0, \sum_{j=0}^{n} x_j = 1 \right\} \quad .$$

A primitive set will consequently consist of $n+1$ vectors with $n+1$ components each. A set $X = \left\{ x^o, x^1, \ldots, x^h \right\}$ of $(n+1)$ dimensional vectors is selected, where $h > n$. The first $n+1$ vectors of A are given by

$$x^o = (0, M_o, M_o, \ldots, M_o)' \quad ,$$
$$x^1 = (M_1, 0, M_1, \ldots, M_1)' \quad ,$$
$$\vdots$$
$$x^n = (M_n, M_n, M_n, \ldots, 0)' \quad ,$$

where $M_0 > M_1 > M_2 > \ldots M_n > 1$, whereas the (h−n) remaining vectors are selected so as to have a rather even
distribution over S and such that no two vectors have the
same i^{th} coordinate for any i . Each vector x^j in X
is associated with the corresponding column in a matrix A
as follows:

$$
A = \begin{array}{c} \begin{array}{cccccccc} x^o & x^1 & x^2 & \ldots x^n & x^{n+1} & \ldots\ldots\ldots x^j & \ldots x^h \end{array} \\ \left[\begin{array}{ccccccc} 1 & 0 & 0\ldots\ldots0 & a_{o,n+1}\cdots\cdots\cdot a_{oj}\cdots\cdot a_{oh} \\ 0 & 1 & 0\ldots\ldots0 & a_{1,n+1}\cdots\cdots\cdot a_{1j}\cdots\cdot a_{1h} \\ 0 & 0 & 1\ldots\ldots0 & a_{2,n+1}\cdots\cdots\cdot a_{2j}\cdots\cdot a_{2h} \\ \cdot & \cdot & \cdot \quad\cdot & \cdot \qquad\qquad\cdot \qquad\cdot \\ \cdot & \cdot & \cdot \quad\cdot & \cdot \qquad\qquad\cdot \qquad\cdot \\ \cdot & \cdot & \cdot \quad\cdot & \cdot \qquad\qquad\cdot \qquad\cdot \\ \cdot & \cdot & \cdot \quad\cdot & \cdot \qquad\qquad\cdot \qquad\cdot \\ 0 & 0 & 0\ldots\ldots1 & a_{n,n+1}\cdots\cdots\cdot a_{nj}\cdots\cdot a_{nh} \end{array} \right] \end{array}
$$

The column a_j with $j > n$ is constructed according
to the following:

<u>Rule of Association</u>: For every x^j , $j > n$ let

    1) $a_j = (1,1+f_1(x^j),\ldots,1+f_n(x^j))'$ if $g_i(x^j) \leq 0$

        for all $i=1,\ldots,m$.

    2) $a_j = (1,1-g_{i1}(x^j),\ldots,1-g_{in}(x^j))'$ if $g_i(x^j) > 0$

        for some i . If the inequality holds for more
        than one i , then select the smallest such i .

      The vector b is taken as $(1,\ldots,1)'$. The con-
straint set $\{y\,|\,y \geq 0,\ Ay = b\}$ is clearly bounded since the
top row of A has all its entries equal to one, aside from

the first (n+1) columns. The combinatorial theorem, stated in §2 may then be applied and we obtain a primitive set $\{x^{j_o}, x^{j_1}, \ldots, x^{j_n}\}$ whose associated columns form a feasible basis for $Ay = b$, i.e.,

$$
y_{j_o}\begin{bmatrix} a_{oj_o} \\ \cdot \\ \cdot \\ \cdot \\ \cdot \\ \cdot \\ \cdot \\ a_{nj_o} \end{bmatrix} + \ldots + y_{j_n}\begin{bmatrix} a_{oj_n} \\ \cdot \\ \cdot \\ \cdot \\ \cdot \\ \cdot \\ \cdot \\ a_{nj_n} \end{bmatrix} = \begin{bmatrix} 1 \\ \cdot \\ \cdot \\ \cdot \\ \cdot \\ \cdot \\ \cdot \\ 1 \end{bmatrix}
\tag{3}
$$

with $y_{j_o}, \ldots, y_{j_n} \geq 0$.

If the grid is sufficiently fine the non-slack vectors in the primitive set represent an approximate stationary point and the weights $y_{j_o}, \ldots, y_{j_n}$ may be used to define the associated Lagrange multipliers. In order to show that for sufficiently large h any of the non-slack vectors x^{j_i} , $j_i > n$, represents an approximation of a stationary point we consider a sequence of finer and finer grids on the simplex. We select a sequence for which all of the non-slack vectors converge to $\hat{x}$, the columns of A in the final feasible bases converge, and the corresponding weights $y_{j_o}, \ldots, y_{j_n}$ converge to $\hat{y}_{j_o}, \ldots, \hat{y}_{j_n}$.

The non-slack columns of A in the final feasible bases will converge to either of the two types:

(i) $(1, 1 + f_1(\hat{x}), \ldots, 1 + f_n(\hat{x}))'$, or

(ii) $(1, 1 - g_{i1}(\hat{x}), \ldots, 1 - g_{in}(\hat{x}))'$.

Relation (3) then takes the limiting form:

$$\hat{\delta} \begin{bmatrix} 1 \\ 1 + f_1(\hat{x}) \\ \cdot \\ \cdot \\ \cdot \\ \cdot \\ 1 + f_n(\hat{x}) \end{bmatrix} + \sum_{i=1}^{m} \hat{\varepsilon}_i \begin{bmatrix} 1 \\ 1 - g_{i1}(\hat{x}) \\ \cdot \\ \cdot \\ \cdot \\ \cdot \\ 1 - g_{in}(\hat{x}) \end{bmatrix} \leq \begin{bmatrix} 1 \\ 1 \\ \cdot \\ \cdot \\ \cdot \\ \cdot \\ 1 \end{bmatrix} \quad (4)$$

where $\hat{\delta}$ and $\hat{\varepsilon}_i$ are sums of limiting weights $\hat{y}_{j_k}$ for the respective cases.

The rules of association then imply:

Equality in the jth row of (4) if $\hat{x}_j > 0$.   (5.1)

If $\hat{\delta} > 0$, then $g_i(\hat{x}) \leq 0$ for all i .   (5.2)

If $\hat{\varepsilon}_i > 0$, then $g_i(\hat{x}) \geq 0$.   (5.3)

We shall demonstrate that $\hat{\delta} > 0$, and the desired result will then follow. To obtain this result we shall require the convexity of the functions $g_i(x)$ and the constraint qualification, i.e., the existence of a vector x^* with $g_i(x^*) < 0$ for all i .

If $\hat{\delta} = 0$ and we multiply the jth inequality in (4) by $(x_j^* - \hat{x}_j)$ $(x_o^* = 1 - \sum_{i=1}^{n} x_i^* > 0)$ we obtain

$$- \sum_{i=1}^{m} \hat{\epsilon}_i \sum_{j=1}^{n} g_{ij} (\hat{x}) \ (x_j^* - \hat{x}_j) \leq 0 \quad . \tag{6}$$

The convexity assumption on the functions $g_i(x)$ implies

$$g_i(x^*) - g_i(\hat{x}) \geq \sum_{j=1}^{n} g_{ij}(\hat{x}) \ (x_j^* - \hat{x}_j) \quad . \tag{7}$$

This inequality is incorporated in (6) and we obtain:

$$\sum_{i=1}^{m} \hat{\epsilon}_i g_i(\hat{x}) \leq \sum_{i=1}^{m} \hat{\epsilon}_i g_i(x^*) \quad . \tag{8}$$

At least one of $\hat{\epsilon}_i$ must be strictly positive. The constraint qualification implies therefore that

$$\sum_{i=1}^{m} \hat{\epsilon}_i g_i(x^*) < 0 \quad ,$$

i.e.,

$$\sum_{i=1}^{m} \hat{\epsilon}_i g_i(\hat{x}) < 0 \quad ,$$

a contradiction of (5.3) We consequently conclude that $\hat{\delta} > 0$.

Knowing that $\hat{\delta} > 0$ we have from (5.1) that $g_i(\hat{x}) \leq 0$ for all i . The boundedness assumption implies that $\hat{x}_1 + \ldots + \hat{x}_n < 1$, i.e., $\hat{x}_o > 0$. We may therefore conclude from (4) that

$$\hat{\delta} + \sum_{i=1}^{m} \hat{\epsilon}_i = 1 \quad .$$

Let

$$\hat{\lambda}_i = \hat{\varepsilon}_i / \hat{\delta} \quad .$$

(4) may then be rewritten as

$$f_j(\hat{x}) - \sum_{i=1}^{m} \hat{\lambda}_i g_{ij}(\hat{x}) \leq 0 \quad , \text{ for all } j \quad , \qquad (9.1)$$

with equality if $\hat{x}_j > 0$.

(5.2) and (5.3) permit us further to state that

$$g_i(\hat{x}) \leq 0 \quad \text{for all } i \quad , \qquad (9.2)$$

with equality if $\hat{\lambda}_i > 0$.

We conclude therefore that $\hat{x}$ and the associated Lagrange multipliers $\hat{\lambda}$ represent a stationary point. Moreover, we have that $\hat{x}$ is the optimal solution to (1) if $f(x)$ is concave.

5. A Procedure for Generating more than one Stationary Point

A primitive set $x^{j_0}, x^{j_1}, \ldots, x^{j_n}$ whose corresponding columns $a_{j_0}, a_{j_1}, \ldots, a_{j_n}$ of A form a feasible basis for $Ay = b$ represents an approximation of a stationary point. If $f(x)$ is not concave the program may have multiple stationary points and we are therefore interested in an algorithm that generates multiple primitive sets of this type, each primitive set representing an approximation of a stationary point.

We shall describe below a procedure for generating multiple primitive sets, with the corresponding columns of each set forming a feasible basis for $Ay = b$.

The combinational algorithm described in §3 started in a specific corner of the simplex. There is no reason we should not start in any corner of the simplex; say the ith corner, with the primitive set consisting of the vectors $x^0, x^1, \ldots, x^{i-1}, x^{i+1}, \ldots, x^n$ and a single vector x^j ($j > n$) . The vector x^j must be that vector in X (other than the first $n+1$ members of X) with the largest i^{th} coordinate. On the other hand the columns $0, 1, \ldots, n$ form a feasible basis for $Ay = b$.

We bring a_j into the basis. If a_i is knocked out of the basis we are through; otherwise, say a_α is knocked out of the basis, x^α is then next replaced in the primitive set. The algorithm proceeds as before, alternating between pivot operations on the system of equations and replacement operations on the primitive set and terminates when either x^i is brought into the primitive set or a_i is knocked out of the basis.

The algorithm may thus be initiated in any of the $n+1$ corners of the simplex.

When we have located a primitive set $x^{j_0}, x^{j_1}, \ldots, x^{j_n}$ whose corresponding columns $a_{j_0}, a_{j_1}, \ldots, a_{j_n}$ of A form a feasible basis for $Ay = b$, we can also initiate the algorithm from this primitive set:

    i) By replacing x^i if x^i is a member of the primitive set. Say the replacement is x^{j_*} .

or

    ii) By bringing a_i into the basis if x^i is not a

        member of the primitive set. If a_{j_α} is then

        knocked out of the basis x^{j_α} is then replaced

        in the primitive set. Say the replacement is

        x^{j_*} .

a_{j_*} is next brought into the basis for the system of equa-

tions and the algorithm proceeds as before and can only stop

when either

    i) a_i is knocked out of the basis or x^i is

        brought into the primitive set. We have then

        located another primitive set whose corre-

        sponding columns in A form a feasible basis

        for the system of equations $Ay = b$,

    or

    ii) The replacement operation on primitive sets can-

        not be performed because the primitive set con-

        sists of the vectors $x^0, x^1, \ldots, x^{i-1}, x^{i+1}, \ldots, x^n$

        and one vector x^j , and we are trying to remove

        the latter. The algorithm has thus taken us to

        the ith corner of the simplex.

There are thus $n+1$ different ways that the algorithm may

be initiated at a primitive set whose corresponding columns

form a feasible basis for $Ay = b$.

    The following figure illustrates in the case of $n=2$ a

possible relationship between the corners of the simplex and

primitive sets whose corresponding columns form a feasible

basis.

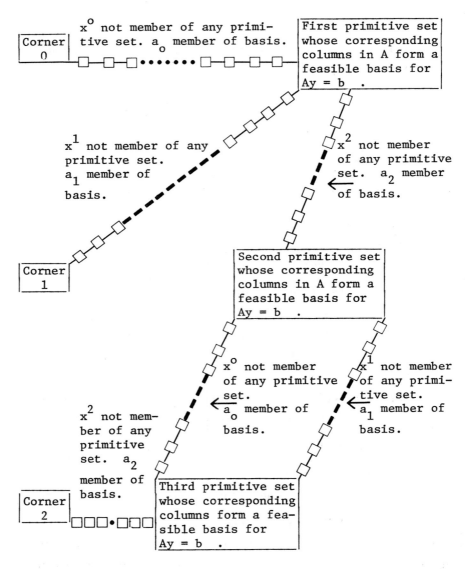

Suppose we initiate the algorithm from all corners of the simplex and take all possible exits from the primitive sets whose corresponding columns in A form a feasible basis for $Ay = b$, but we avoid repeating sequences. Let K

designate the number of different primitive sets with the
desired property that are found. We shall then have to run
the algorithm $((K+1)/2) \cdot (n+1)$ times (as always in these
kinds of problems K is odd). As the above figure illustrates, when n is 2 and K is 3 , there are six different sequences of primitive sets.

　　　We would like to stress that the procedure described
above is not guaranteed to locate all primitive sets whose
corresponding columns in A form a feasible basis for
Ay = b since primitive sets with this property may be connected as illustrated in the following figure in the case of
n = 2 .

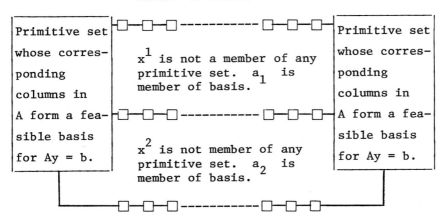

The algorithm is tested on three examples of the following type:

$$\text{maximize} \quad f(x) = \sum_{j=1}^{4} c_j x_j + \frac{1}{2} \sum_{j=1}^{4} \sum_{k=1}^{4} d_{jk} x_j x_k$$

$$g_i(x) = \sum_{j=1}^{4} t_{ij}x_j - r_i \leq 0 \quad , \quad i=1,\ldots,4$$

subject to

$$x \geq 0 \quad .$$

All examples had 3 stationary points. The procedure described located all stationary points. In one of the examples the same stationary point was approximated by 3 different primitive sets. This example illustrates the fact that the number of primitive sets, whose corresponding collumns form a feasible basis for $Ay = b$, may exceed the number of stationary points.

REFERENCES

[1] Eaves, B. C.; "Nonlinear Programming via Kakutani Fixed
     Points", Working Paper No. 294, Center for Operations
     Research in Management Science, University of Califor-
     nia, Berkeley, 1970.

[2] Hansen, T., "A Fixed Point Algorithm for Approximating
     the Optimal Solution of a Concave Programming Problem",
     Cowles Foundation Discussion Paper No. 277, 1969.

[3] Lemke, C. E., and Howson, J. T., "Equilibrium Points of
     Bi-Matrix Games", SIAM Journal of Applied Mathematics,
     12, 1964.

[4] Merrill, O. H., "Applications and Extensions of an
     Algorithm that Computes Fixed Points of Certain Non-
     Empty Convex Upper Semi-Continuous Point to Set
     Mappings", Technical Report no. 71-7, Department of
     Industrial Engineering, University of Michigan, 1971.

[5] Scarf, H. with the collaboration of T. Hansen, "The
     Computation of Economic Equilibria", Yale University
     Press, 1973.

A Method of Continuation for
Calculating a Brouwer Fixed Point

R. B. Kellogg, T. Y. Li, J. Yorke

1.   Introduction

Let D be a bounded open convex domain in R^n , and
let $F : \overline{D} \to \overline{D}$ be continuous. The Brouwer Fixed Point Theorem guarantees the existence of a fixed point, a point $x^o \in \overline{D}$
such that $F(x^o) = x^o$. The broad applicability of this
theorem makes the associated numerical problem important,
namely "give a generally applicable algorithm for finding the
fixed point". A numerical method which can be implemented by
computer was first found by Scarf [1]. The approach of Scarf
is based in part on the use of Sperner's Lemma method in
proving the Brouwer Fixed Point Theorem. The domain D (assumed to be a simplex) is simplicially decomposed and the
systematic search procedure produces a computationally feasible way to find an "approximate" fixed point of F . This
method has been extended and simplified in a number of ways.
(See the other papers in this volume.)

In this paper we introduce a different method for computing a fixed point of a C^1 function of F . Our algorithm is modeled on a different proof of the Brouwer theorem,
given in [2,3]. In this proof one reduces the existence of a
fixed point to the non-existence of a retraction H of $\overline{D}$
onto ∂D . The retraction is shown not to exist by assuming
that it does and studying the curves $H^{-1}(x^o)$, $x^o \in \partial D$. To
formulate our algorithm, we construct a retraction
$H : \overline{D} \setminus C \to \partial D$, where C is the set of fixed points. We

then show that the curves $H^{-1}(x^o)$ lead from a point

$x^o \in \partial D$ to the set C . The algorithm involves following

these curves.

We state the main result in §2 and present the algorithm in §3. Some additional results and discussion are contained in §4. The proofs of the theorems and numerical results will be given in a subsequent paper.

2. Main Result

Let D be a bounded open convex set in R^n , let ∂D
denote its boundary. We assume that ∂D consists of a finite number of smooth hypersurfaces. Let $F : \overline{D} \to \overline{D}$ be a
C^1 function. Let $C = \{x | F(x) = x\}$ be the set of fixed
points of F . For $x \in \overline{D} \setminus C$ we define $H(x)$ to be the
point of intersection of ∂D with the ray drawn from $F(x)$
through x . Notice that $H(x) = x$ for $x \in \partial D$. Also
note that $H(x)$ is of class C^1 at points x such that
$H(x)$ lies on a smooth part of ∂D . The main theorem, on
which the algorithm is based, is as follows.

Theorem 1. For almost every $x^o \in \partial D$ (in the sense of

$(n-1)$-dimensional measure), there is an open set $U \supset H^{-1}(x^o)$

such that $H'(x)$ is of rank n-1 for $x \in U$. Also, the

connected component of the set $H^{-1}(x^o)$ which contains x^o

is a curve $x(t), 0 < t < T \leq \infty$. Furthermore, for each

$\varepsilon > 0$ there is a $t(\varepsilon) < T$ such that for $t(\varepsilon) < t < T$,

dist. $(x(t),C) < \varepsilon$.

The curve $x(t)$ given by Theorem 1 is called the continuation curve starting at x^o . It satisfies the equations

$$
\begin{cases}
H(x(t)) = x^o \; , \\
\\
x(0) = x^o \; .
\end{cases}
\tag{1}
$$

The proof of Theorem 1, which is in essence given in [2], involves a use of the implicit function theorem. Since $H^1(x^o)$ has rank $n-1$, the implicit function theorem guarantees the existence of a neighborhood U of $x^o = H(x^o)$ such that the set $H^{-1}(x^o) \cap U$ consists of a curve $x(t)$, with $x(0) = x^o$. This starts the construction of the continuation curve. Repeated use of the implicit function theorem insures that the curve can be continued and does not cross itself. A compactness argument shows that $x(t)$ tends to the set C of fixed points. As mentioned in §1 , the proof of Theorem 1 is essentially constructive and can be used to prove the Brower fixed point theorem.

3. The Algorithm

In this section we let D be the unit cube of R^n with one face lying on $x_1 = 0$. The continuation curve of Theorem 1 may be given by an ordinary differential equation. Setting $dx/dt = \dot{x}$, we get from (1) $H'(x)\dot{x} = 0$. By Theorem 1 for almost all $x^o \in \partial D$, $H'(x)$ for x on $x(t)$

always has rank   n-1   .   Hence, for   x   near   x(t)   , there

exist $z(x) \neq 0$ such that

$$H'(x)z(x) = 0 \qquad\qquad (2)$$

and the curve   x(t)   is given by the equation

$$\dot{x}(t) = z(x(t)), \quad x(0) = x^o \quad . \qquad\qquad (3)$$

If   t   represents arc length, we have, in addition,

$$|z(x(t))|^2 = 1 \quad . \qquad\qquad (4)$$

Let $x^o \in \partial D$ with $x_1^o = 0$.   Then   H(x)   for   x   on   x(t)

can be written as

$$H(x) = x - \frac{x_1}{x_1 - F(x)} (x - F(x)) \qquad\qquad (5)$$

and   H'(x)   for   x   on   x(t)   is given by the following

matrix:

$$H'(x) = \begin{bmatrix} 0 & \cdot & \cdot & \cdot & 0 \\ \dfrac{\partial H_2}{\partial x_1} & \cdot & \cdot & \cdot & \dfrac{\partial H_2}{\partial x_n} \\ \cdot & & & & \cdot \\ \cdot & & & & \cdot \\ \cdot & & & & \cdot \\ \dfrac{\partial H_n}{\partial x_1} & & & & \dfrac{\partial H_n}{\partial x_n} \end{bmatrix}$$

Let

$$h_i(x) = \left(\frac{\partial H_i}{\partial x_1}, \ldots, \frac{\partial H_i}{\partial x_n} \right), \quad 2 \le i \le n \quad . \quad (6)$$

Since $H'(x)$ has rank $n-1$, all $h_i(x)$, $2 \le i \le n$, are linearly independent. From (2) we see that

$$z(x) \perp h_i(x), \quad 2 \le i \le n \quad ,$$

for all x on $x(t)$.

Summarizing the above observations, we give the following algorithm. Here we suppose ε is chosen as a small positive number.

Step 1. Choose $x^o \in \partial D$ with $x_1^o = 0$ and compute $H'(x^o)$.
The matrix $H'(x^o)$ is given by

$$H'(x^o) = \begin{bmatrix} 0 & 0 & \cdot & \cdot & \cdot & 0 \\ \dfrac{\partial H_2}{\partial x_1} & 1 & \cdot & \cdot & \cdot & 0 \\ \cdot & & & & & \\ \cdot & & & \cdot & & \\ \cdot & & & & \cdot & \\ \dfrac{\partial H_n}{\partial x_1} & 0 & & & & 1 \end{bmatrix}$$

Letting $s = \left(1 + (\partial H_2/\partial x_1)^2 + \ldots + (\partial H_n/\partial x_1)^2\right)^{1/2}$, we may write $z(x^o)$ as

$$(z(x^o))_1 = -1/s$$

$$(z(x^o))_i = (\partial H_i/\partial x_1)/s, \qquad 2 \le i \le n .$$

Step 2. Compute $x^1 = x^o + \delta_o z(x^o)$ (Euler method [4] with step size δ_o (see remark 1)) and calculate $|F(x^1) - x^1|$. Terminate the algorithm if $|F(x^1) - x^1| < \varepsilon$.

Step 3. Suppose we have calculated x^k by

$$x^k = x^{k-1} + \delta_{k-1} z(x^{k-1}) .$$

We then compute $H'(x^k)$ and $h_i(x^k)$ for $2 \le i \le n$ as defined by (6). We shall write h_i for $h_i(x^k)$ when no clarification is needed. Let L_k be the linear space spanned by the $n-1$ linearly independent vectors $h_2, \ldots, h_n$. Since $z(x^k)$ is perpendicular to L_k and if δ_{k-1} is small, $z(x^{k-1}) \notin L_k$, we may write

$$z(x^k) = z(x^{k-1}) + \sum_{i=2}^{n} a_i h_i \quad . \tag{7}$$

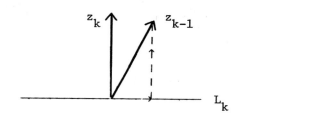

By taking the inner product with h_j for $2 \le j \le n$ on both sides of (7), we have

$$0 = (z(x^k), h_j) = (z(x^{k-1}), h_j) + \sum_{i=2}^{n} a_i(h_i, h_j) \quad .$$

That is

$$\sum_{i=1}^{n} a_i(h_i, h_j) = -(z(x^{k-1}), h_j), \quad 2 \le j \le n \quad . \tag{8}$$

Since all h_i , $2 \le i \le n$, are linearly independent, the

matrix (h_i, h_j) is a positive definite symmetric matrix.
We solve (8) for a_i, $2 \leq i \leq n$, and then calculate $z(x^k)$
by (7). Choosing step size δ_k we write

$$x^{k+1} = x^k + \delta_k (z(x^k)/|z(x^k)|) \quad .$$

Step 4. Calculate $|F(x^{k+1}) - x^{k+1}|$ and terminate the algo-
rithm if $|F(x^{k+1}) - x^{k+1}| < \varepsilon$.

 The following remark gives some criteria for choosing
step size δ_k .

Remark 1. Using uniform step size on Euler's method to approximate the solution of (3), we can not usually have a
satisfactory termination of the algorithm. That is, because
we approximate singularities of the differential equation (3)
when we are close to the fixed points of F . We give here
two criteria for automatically adjusting the step size of
each preceeding step. First of all, let δ_o be the initial
step size and $\{\varepsilon_k\}$ be a sequence of positive numbers with
$d = \sum_0^\infty \varepsilon_k < \infty$, and the $(n-1)$-neighborhood of x^o with
radius d on the hyperplane $x_1 = 0$ is contained in ∂D .
At x^k , we may set $\delta_k = \delta_{k-1}$ as a temporary step size to
calculate

$$y^{k+1} = x^k + \delta_k z(x^k) \quad . \tag{9}$$

If

$$|H(y^{k+1}) - H(x^k)| < \varepsilon_k \qquad (10)$$

we let $x^{k+1} = y^{k+1}$. Otherwise we redefine $\delta_k = \delta_{k-1}/2$ and recalculate y^{k+1}. We repeat this process until (10) is satisfied. Secondly, since $F_1(x^o) > x_1^o$, if there exists t with $F_1(x(t)) < (x(t))_1$, then we have $F_1(x(t_o)) = (x(t_o))_1$ for some $t_o < t$. This is impossible since $x(t)$ is characterized by (5). Therefore, when we have calculated (9) we may calculate $F_1(y^{k+1}) - (y^{k+1})_1$. If

$$F_1(y^{k+1}) - (y^{k+1})_1 < 0 , \qquad (11)$$

we cut the step size in half and recalculate y^{k+1} until the reverse inequality in (11) is satisfied and then let $x^{k+1} = y^{k+1}$.

Remark 2. If $S \subset R^n$ denotes the $(n-1)$-dimensional simplex

$$S = \{x \mid \sum_1^n x_i = 1, x_i \geq 0\} ,$$

we let $T \subset R^{n-1}$ be the set given by

$$T = \{\overline{x} = (x_1,\ldots,x_{n-1}) \mid \sum_1^{n-1} x_i \leq 1, \; x_i \geq 0\} \quad .$$

We define $\overline{F} : T \to T$ by

$$\overline{F}(\overline{x}) = (\overline{F}_1(\overline{x}),\ldots,\overline{F}_{n-1}(\overline{x})),$$

where
$$\overline{F}_i(\overline{x}) = F_i(x_1,\ldots,x_{n-1}, \; 1 - \sum_1^{n-1} x_i) \quad .$$

Then since one face of the convex set T lies on $x_1 = 0$, we may use the algorithm as described above to obtain a fixed point of $\overline{F}$, and hence of F .

4. Additional Results

In this section we give several additional results which complement Theorem 1. We start by considering when $H'(x)$ has rank $n-1$. Since $H(x)$ lies on the ray from $F(x)$ through x , we may write

$$H(x) = (1 - \mu(x))x + \mu(x)F(x)$$

where $\mu:\overline{D}\backslash C \to R^1$ is given by the conditions

$$\mu(x) < 0$$

$$H(x) \in \partial D \quad .$$

The matrix $H'(x)$ may be described by its action on a vector v as follows:

$$H'(x)v = [(1-\mu(x))I + \mu(x)F'(x)]v$$

$$+ (\mu'(x), v)(F(x) - x) .$$

Setting $A(x) = (1-\mu(x))I + \mu(x)F'(x)$, we see that $H'(x)$ is a perturbation of rank 1 of $A(x)$. From this it follows that if $A(x)$ is non-singular, then $H(x)$ has rank $n-1$ and

$$u(x) = A(x)^{-1}[F(x)-x] \tag{12}$$

is a null vector of $H'(x)$. This leads to

Theorem 2. Let the conditions of Theorem 1 be satisfied, and in addition suppose that for each $x \in D$, $F'(x)$ has no eigenvalues in the interval $(1,\infty)$. Then the continuation curve may be started at any x^{o} lying in a smooth portion of ∂D .

The hypotheses of Theorem 2, and the formula (12), leads to a relation between the continuation method presented here and the continuous Newton's method [5, p. 744]. Using (12), the differential equation (3) may be written

$$\frac{dx}{ds} = -[F'(x) - I + \mu(x)^{-1}I]^{-1}(F(x) - x) . \tag{13}$$

The parameter s of (13) is no longer arc length. As $x(s)$ approaches the set C, it is easily seen that $\mu(x(s) \to \infty$. Consequently, (13) is similar to the differential equation of the continuous Newton's method,

$$\frac{dx}{ds} = -[F'(x) - I]^{-1}(F(x) - x)$$

for solving the equation $F(x) - x = 0$.

The continuation method presented here may be applied to solving equations of the form $G(x) = 0$ where $G:R^n \to R^n$ is an M-function [6,7] of class C^1. It may be shown that for $\alpha > 0$ sufficiently small, the function $F(x) = x - \alpha G(x)$ maps a suitably chosen n-cube into itself. The continuation method, applied to the function F, gives rise to a continuation method for finding the solution x^* of

$$G(x^*) = 0 . \tag{14}$$

To give this method, let x^o be a subsolution of (14); that is, x^o is a vector satisfying $G(x^o) < 0$. (As is customary in the theory of M-function, the vector inequality $u < v$ means the corresponding inequality for each of the components, $u_i < v_i, 1 \le i \le n$.) We define a function K by

$$K(x) = x - \mu(x)G(x) ,$$

$$\mu(x) = \frac{x_1 - x_1^o}{G_1(x)} .$$

The first component of $K(x)$ satisfies $K_1(x) = x_1^o$. Thus the matrix $K'(x)$ is singular. We have the following theorem.

<u>Theorem 3</u>. For each x such that $G_1(x) < 0$, $K'(x)$ has rank $n-1$.

$$K(x(t)) = x^1$$

$$(15)$$

$$x(0) = x^o$$

define a curve $x(t)$, $0 \leq t < T \leq \infty$, with the following properties ($\cdot = d/dt$) :

$$\dot{x}(t) > 0 , \qquad 0 \leq t < T$$
$$G(x(t)) < 0 , \qquad 0 \leq t < T$$
$$\mu(x(t)) < 0 , \qquad 0 \leq t < T$$
$$\dot{\mu}(x(t)) < 0 , \qquad 0 \leq t < T$$
$$\lim_{t \to T} x(t) = x^* .$$

This theorem may be used to devise algorithms for solving equation (14).

---
*Research supported in part by National Science Foundation Grant GP-20555.

**Research supported in part by National Science Foundation Grant GP-31386X.

REFERENCES

[1] Scarf, H. "The Approximation of Fixed Points of a Con-
     tinuous Mapping", SIAM Journal of Applied Mathematics,
     15, 5 (1967), 1328-1343.

[2] Milnor, J. W. Topology From the Differentiable View-
     point, University Press of Virginia, Charlottesville,
     1965.

[3] Hirsch, M. "A Proof of the Nonretractibility of a Cell
     onto its Boundary", Proc. Amer. Math. Soc. 14 (1963),
     364-365.

[4] Isaacson, E., and Keller, H. B. Analysis of Numerical
     Methods, Wiley, New York, 1966, esp. p. 367 et seq.

[5] Meyer, G. H. "On Solving Nonlinear Equations with a One-
     Parameter Operator Imbedding", SIAM J. Numer. Anal.,
     Vol. 5 (1968), 739-752.

[6] Rheinboldt, W. "On M-Fundtions and Their Application to
     Nonlinear Gauss-Seidel Iterations and Network Flows",
     J. Math. Anal. Appl., 32 (1970), 274-307.

[7] Rheinboldt, W. "On Classes of n-dimensional Nonlinear
     Mappings Generalizing Several Types of Matrices" in
     Numerical Solution of Partial Differential Equations-II,
     B. Hubbard, ed., Academic Press, 1971.

Continuation Methods for Simplicial Mappings

C. B. Garcia

ABSTRACT

Let D be an open bounded set in E^n and $F: \overline{D} \to E^n$ a continuous mapping. It is shown that for any $y \in D$, and any nonsingular A there exists a connected compact set $C \subseteq \overline{D}$ containing y such that $(1-t)A(x-y) + t\,Fx = 0$, for some $t \in [0,1]$ and for all $x \in C$; moreover if $C \subseteq D$ then F has a zero in C.

It is then shown how the result may be applied for the fixed point problem and the nonlinear complementarity problem.

In this paper, we consider the model

$$H: E^n \times [0,1] \to E^n \ , \tag{1}$$

where H is a continuous mapping such that

$$H(x,0) = \overset{o}{F} x \ , \ H(x,1) = Fx \ , \quad x \in E^n \ . \tag{2}$$

A solution x^0 of $F^0 x = 0$ is known, and the equation
$Fx = 0$ is to be solved.

Under reasonable conditions on F , it may be shown
that there is a solution curve $p: [0,1] \to E^n$ which depends
continuously on t such that

$$H(p(t),t) = 0 \ , \ \forall t \in [0,1] \ . \tag{3}$$

Methods which move along this solution curve are known as
continuation methods. The basic idea of the method is to use
a locally convergent iterative process for a stepwise approximation to the solution curve (see, e.g., [13]; see also the
paper of Li and Kellogg which appears in this volume). Recently, constructive methods for approximating the solution
curve of H have been formulated by introducing a simplicial
subdivision of $E^n \times [0,1]$, and by approximating F^0 and F
by continuous simplicial mappings. (A partial list of works
includes [1, 2, 6, 8, 12]. The essence of these methods is

based on the ideas of Lemke and Howson [9, 11], and Scarf
[14, 15].) Theorem 1 of this paper states that such a solution curve exists for continuous simplicial mappings. In the
limiting case, we get existence theorems asserting existence
of a set C "connecting" the known solution $(\overset{o}{x},0)$ of
$H(x,0) = 0$ to an unknown solution of $H(x,1) = 0$, and
where $H(x,t) = 0$ for some $t \in [0,1]$, for all $x \in C$.
See Eaves [1] for a "homotopy principle" for complementarity
problems. Other papers of interest are [3, 4, 5, 7, 10, 16].

1. Definitions

Let K be a simplicial subdivision (or triangulation)
of E^n , and T a simplicial subdivision of $E^n \times [0,1]$
such that (i) $E^n_i = E^n \times \{i\}$ for $i = 0,1$, have the same
structure as K , and (ii) all 0-simplices of T lie in
E^n_i , $i = 0,1$. By a triangulation T^δ of $E^n \times [0,1]$,
$\delta > 0$, we mean a triangulation where the mesh size of K^δ
is less than δ .

Given a triangulation T , let $\phi^i : E^n \to E^n$, $i = 0,1$
be continuous simplicial mappings, i.e.,

$$\phi^i x = \sum_{j=0}^{n} \lambda_j \phi^i v^j$$

whenever   (4)

$$(x,i) = \sum_{j=0}^{n} \lambda_j (v^j,i) \ , \ \lambda \geq 0 \ , \ \sum_{j=0}^{n} \lambda_j = 1$$

for some n-simplex $\sigma = \{(v^j, i)\}_{j=0}^{n}$ of E_i^n .

Denote by e^j the n-vector with 1 in the jth component, 0 elsewhere, and by e the n-vector of ones. For a set C , let $|C|$ denote the cardinality of C , $\overline{C}$ its closure, ∂C its boundary, and int C its interior.

2. Main Results

Let T be a triangulation of $E^n \times [0,1]$ and $\phi^i : E^n \to E^n$ be continuous simplicial mappings. Define $\Phi : E^n \times [0,1] \to E^n$ by

$$\Phi(x,t) = (1 - t)\phi^0 w + t\phi^1 z , \quad \forall(x,t) \in E^n \times [0,1] \quad , \quad (5)$$

where w and z are points of E^n satisfying

(i) $x = (1 - t) w + tz$

(ii) (x,t) , $(w,0)$, and $(z,1)$ all lie in the same (n+1)-simplex of T . Clearly, Φ defined in this way is of the form (1).

The following theorem may be shown through the use of Merrill's algorithm [12].

A point $(x,0) \in E_o^n$ is said to be regular if it is interior to an n-simplex of E_o^n .

__Theorem 1.__ Given a triangulation T of $E^n \times [0,1]$ and continuous simplicial mappings $\phi^i : E^n \to E^n$, suppose

    (i) $\overset{o}{\phi}x = 0$ for an odd number of $(x,0) \in E_o^n$, and

    (ii) $\overset{o}{\phi}x = 0$ implies $(x,0)$ is regular.   (6)

Then, there exists a continuous path $p: [0,a) \to E^n \times [0,1]$
such that $\Phi \cdot p(\theta) = 0$ for $0 \le \theta < a$, where Φ is defined
by (5). Furthermore, let $p(\theta) = (p_o(\theta), p_1(\theta))$ where
$p(\theta) \in E^n$, $p_1(\theta) \in [0,1]$. Then

    (i) $p_1(0) = 0$;

    (ii) $0 < p_1(\theta) < 1$, for each $0 < \theta < a$; and

    (iii) If $a = \infty$, then $p[0,\infty)$ is an unbounded   (7)
            path; if $a < \infty$, then $p(a) = \lim\limits_{\theta \to a^-} p(\theta)$

            exists, and $\Phi \cdot p(a) = 0$, $p_1(a) = 1$.

Let us now turn to the problem of finding a zero of a continuous function F .

<u>Theorem 2.</u> Let D be an open bounded set in E^n and
$F: \overline{D} \to E^n$ a continuous mapping. For any $y \in D$, and any
nonsingular nxn matrix A there exists a connected compact
set $C \subseteq \overline{D}$ containing y and

$$H(x,t) \overset{\Delta}{=} (1-t)A(x-y) + t\,Fx = 0 \text{ , for some } t \in [0,1] \text{ ,}$$
$$\forall x \in C \text{ .} \quad (8)$$

Moreover if $C \subseteq D$ then $Fx = 0$ for some $x \in C$.

<u>Proof:</u> Given $y \in D$, extend F to the whole space by
letting

$$Fx = A(x-y) \qquad x \notin \overline{D} \quad .$$

Also, let   (9)

$$F^o x = A(x-y) \qquad x \in E^n \quad .$$

For any triangulation T^δ of $E^n \times [0,1]$ let $\phi_\delta^i : E^n \to E^n$ be continuous simplicial mappings where

$$\phi_\delta^i v = F^i v \qquad \text{for each 0-simplex} \qquad (v,i) \in T^\delta \quad .$$

(Here, F^1 is the mapping F).

Now, if $\{\delta_k\}_1^\infty$ is a sequence of positive numbers with $\delta_k \to 0$, let T^{δ_k} be a sequence of triangulations of $E^n \times [0,1]$ such that $(y,0)$ is a regular point of T^{δ_k} for each k . Observe that $(y,0)$ is a unique zero of $\phi_{\delta_k}^o$ for each k . Hence, by Theorem 1, to each δ_k there exists a path

$$p^k : [0,a^k) \to E^n \times [0,1]$$

satisfying (7). Furthermore, $a^k < \infty$ for each k , as $x \notin \overline{D}$ implies $F^i x = A(x-y)$ which implies $\Phi(x,t) \neq 0$ for any $0 \leq t \leq 1$, and any x sufficiently far from $\overline{D}$. Letting

$$C = \overline{\lim_{k \to \infty}} \; p_o^k [0,a^k]$$

(where $\overline{\lim}$ is the set of cluster points), we have

$p^k(0) = (y,0)$ and $\Phi \cdot p^k(\theta) = 0$, $0 \leq \theta \leq a^k$, for each k .

Inasmuch as $\delta_k \to 0$, $\phi_{\delta_k}^i$ are continuous, and $p^k[0,a^k]$ are bounded curves which connect at $(y,0)$, C must be a connected compact set containing y which satisfies (8). It is now apparent that $C \subseteq \overline{D}$ as (8) cannot hold for $x \notin \overline{D}$, any $t \in [0,1]$. Moreover, if $C \subseteq D$ then $p^k(a^k) \in D \times \{1\}$ for an infinite number of k . If $(\overline{x},1)$ is a cluster point of this set, $\overline{x} \in C$ and

$$F\overline{x} = 0 \quad .$$

This proves the theorem.

In the above theorem, D is allowed to be an unbounded set. In this case, the set C is a closed connected set, $C \subseteq \overline{D}$, and when C is compact, $C \subseteq D$ and F has a zero in C .

Theorem 3. Let D be an open bounded set in E^n and $F: \overline{D} \to E^n$ a continuous mapping. For any $y \in D$ and any nonsingular A , there exists a connected compact set $C \subseteq \overline{D}$ containing y such that, for each $x \in C$ there is an index j , $1 \leq j \leq n$ for which $(Ax-Ay)_k \geq 0$, $F_k x \geq 0$, and $(Ax-Ay)_k F_k x = 0$ for $k < j$ and either

(i) $(Ax-Ay)_j \geq 0$; $F_j x \leq 0$; $(Ax-Ay)_k = 0$, $\forall k > j$,

   or   (10)

(ii) $F_j x \geq 0$; $(Ax-Ay)_j \leq 0$; $F_k x = 0$, $\forall k > j$.

If $C \subseteq D$, then $fx = 0$ for some $x \in C$.

Proof: Define $F^i : E^n \to E^n$ as in (9). Let $\phi_\delta^i : E^n \to E^n$ be continuous simplicial mappings defined by

$$\phi_\delta^i v = \begin{cases} e^j & \text{if } j = \min \{k | F_k^i v < 0\} \\ \\ -e & \text{if } F^i v \geq 0 \end{cases} \qquad \begin{array}{l} \text{for each 0-simplex} \\ (v,i) \text{ of } T^\delta . \end{array}$$

With $\{\delta_k\}_1^\infty$, T^{δ_k} defined as in Theorem 1, and

$$C = \overline{\lim_{k \to \infty}} \; p_0^k [0, a^k]$$

we have $C \subseteq \overline{D}$ and $y \in C$. By choice of the $\phi_{\delta_k}^i$, to each $x \in C$ there is an

$$I_o \subseteq \{0,1,2,\ldots,n\}$$

and

$$I_1 = \{0,1,2,\ldots,n\} \sim I_o$$

such that $F_j^i x \leq 0$ and $F_k^i x \geq 0$, $\forall k < j$ if $j \in I_i$, and $F^i x \geq 0$ if $0 \in I_i$.

First, suppose $0 \in I_o$. In this case, let $j = \max \{k | k \in I_1\}$. Then $F_k x = F_k^1 x \geq 0$ for each $1 \leq k < j$ and $F^o x = A(x-y) \geq 0$. Furthermore, for $k < j$, $F_k^i x \leq 0$ where $k \in I_i$, which implies $(Ax-Ay)_k F_k x = 0$. For $k > j$, we have $k \in I_o$ and thus $(Ax-Ay)_k = 0$. Hence (10.i) holds.

Similarly, if $0 \in I_1$, one may show that (10.ii)

holds for this $x \in C$.

Finally, if $C \subseteq D$ then $\Phi \cdot p^k(a^k) = \phi^1_{\delta_k} p^k_o(a^k) = 0$

for an infinite number of k's , which implies that a cluster

point $\bar{x}$ of this sequence is a zero of F .

3. Applications

In this section, we show how the theorems in the previous section may be useful in proving an extension of the Leray-Schauder Theorem (Corollary 4), a theorem of Gould and Tolle (Corollary t), and in finding an approximation to a complementary solution of a nonlinear complementarity problem (Corollary 6).

Corollary 4.   (Extension of the Leray-Schauder Theorem [13])

Let D be an open bounded set in E^n and $f: \overline{D} \to E^n$ be continuous. If for some $y \in D$,

$$fx \neq \lambda x + (1 - \lambda)y \qquad \text{whenever} \qquad \lambda > 1 \qquad \text{and} \qquad x \in \partial D$$

then f has a fixed point in $\overline{D}$.

Proof: Let $F: \overline{D} \to E^n$ be defined by

$$Fx = x - fx \; , \; x \in \overline{D} \; .$$

Then $fx \neq \lambda x + (1 - \lambda)y$ for each $\lambda > 1$, $x \in \partial D$ implies

$$tFx + (1-t)(x-y) \neq 0 \quad \text{for each} \quad 0 \leq t < 1 \; , \; x \in \partial D \; . \quad (11)$$

Now, let C be the connected compact set described in

Theorem 2. If for some $x \in C$ we have $x \in \partial D$, then (8) and (11) imply that $Fx = 0$. Otherwise, $C \subseteq D$ whence F has a zero in C.

Hence, in both cases f has a fixed point in $\overline{D}$.

Corollary 5. (Gould and Tolle [6])

Let D be an open bounded set in E^n and $F: \overline{D} \to E^n$ a continuous mapping. If for some $y \in D$ at least one of the following holds for each $x \in \partial D \cap (\bigcup\limits_1^n D_j)$ where

$$D_j = \{x \in E^n \mid (x_k - y_k)F_k x = 0 , k \neq j\} :$$

     (i) $(x_j - y_j)F_j x > 0$

    (ii) there are indices j and k such that
$$x_j - y_j < 0 \quad \text{and} \quad F_k x < 0$$

  (iii) there are indices j and k, $j < k$ such that
$$x_j - y_j < 0 , F_k x > 0$$

  (iv) there are indices j and k, $j < k$ such that
$$x_k - y_k > 0 , F_j x < 0 .$$

Then F has a zero in D.

Proof: In all cases, one may easily show that (10.i) or (10.ii) does not hold for each $x \in \partial D \cap (\bigcup\limits_1^n D_j)$. Hence, Theorem 3 yields a zero in $C \subseteq D$.

As another application of Theorem 2, consider the problem of finding a solution x to the complementarity problem

$$x \geq 0 , fx \geq 0 , x \, fx = 0 \qquad (12)$$

where $f: E^n \to E^n$ is continuous.

Recall that if $f\bar{x} > 0$ for some $\bar{x} \geq 0$ and f is monotone in E^n_+, then (12) has a solution. (f is monotone in E^n_+ if $(x - y)(fx - fy) \geq 0$ for each $x,y \in E^n_+$. Note that the Kuhn-Tucker conditions of a differentiable and convex nonlinear program $\{\min g_o(x) | g(x) \geq 0 , x \geq 0\}$ gives rise to a complementarity problem with a monotone f.)

__Corollary 6.__ Let $f: E^n \to E^n$ be continuous. Suppose there is an $\bar{x} \geq 0$ such that $f\bar{x} \geq 0$ and $(\bar{x} - x)(f\bar{x} - fx) \geq 0$ $\forall x \in E^n_+$. Then for any scalar $\delta > 0$, there is an $\hat{x} \geq 0$ such that

$$-\delta \leq \min \{\hat{x}_i, f_i\hat{x}\} \leq \delta \qquad \text{for each} \quad i = 1,2,\ldots,n .$$

__Proof:__ Let $F: E^n \to E^n$ be the continuous mapping defined by

$$F_i x = \min \{x_i, f_i x\} \qquad i = 1,2,\ldots,n \qquad \text{for any} \quad x \in E^n .$$

Observe that a zero of F is a solution of (12) and vice versa. We now show that for any $\delta > 0$ there is an $\hat{x} \geq 0$ such that $|F_i\hat{x}| \leq \delta$ for each i .

If $\bar{x}f\bar{x} = 0$, then $\bar{x}$ is a solution to (12) and hence the corollary follows trivially. Thus, assume $\bar{x}f\bar{x} > 0$.

For any $r > 0$ let D be the open ball $B(\bar{x},r)$ in E^n. By Theorem 2, there is a connected compact set $C \subseteq \bar{D}$ containing $\bar{x}$ such that

$$(1-t)(x - \bar{x}) + t \, Fx = 0 \quad \text{for some} \quad t \in [0,1] \text{ , } \forall x \in C \text{ .}$$

If $C \subseteq D$, then F has a zero in C and Corollary 6 holds. Hence, suppose for every $r > 0$ there is an $x^r \in \partial B(\bar{x}, r)$ such that for some $t^r \in (0,1)$

$$Fx^r = \frac{t^r - 1}{t^r} (x^r - \bar{x}) \tag{13}$$

If for some i , $x_i^r < 0$, then from (13) $F_i \, x^r > 0$, which contradicts the choice of F . Hence $x^r \geq 0$.

Now let

$$I = \{i \mid x_i^r \leq \bar{x}_i \text{ , } f_i x^r \leq f_i \, \bar{x}\}$$

$$J = \{i \mid x_i^r > \bar{x}_i\} \text{ .}$$

First, note that (13) yields $f_i x^r \geq 0$ if and only if $x_i^r \leq \bar{x}_i$. Using the monotonicity of f at $\bar{x}$, we have

$$0 \leq (\bar{x} - x^r)(f\bar{x} - fx^r)$$

$$\leq \sum_I (\bar{x}_i - x_i^r)(f_i \bar{x} - f_i x^r) + \sum_J \frac{t^r - 1}{t^r} (\bar{x}_i - x_i^r)^2 \text{ .}$$

But for $i \in I$ we have $0 \leq \bar{x}_i - x_i^r \leq \bar{x}_i$, $0 \leq f_i \bar{x} - f_i x^r \leq f_i \bar{x}$; hence

$$\sum_{I} (\bar{x}_i - x_i^r)(f_i \bar{x} - f_i x^r) \leq \bar{x} \ f\bar{x} \ .$$

We now have

$$\frac{t^r - 1}{t^r} \geq \frac{-\bar{x} \ f\bar{x}}{\sum_{J} (\bar{x}_i - x_i^r)^2} \ .$$

Thus, (13) yields

$$0 \geq F_i x^r \geq \frac{-\bar{x}f\bar{x} \ (x_i^r - \bar{x}_i)}{\sum_{J}(x_i^r - \bar{x}_i)^2} \qquad \text{if} \qquad x_i^r \geq \bar{x}_i$$

$$0 \leq F_i x^r \leq \frac{-\bar{x}f\bar{x} \ (x_i^r - \bar{x}_i)}{\sum_{J}(x_i^r - \bar{x}_i)^2} \qquad \text{if} \qquad x_i^r \leq \bar{x}_i$$

The right-hand sides of the above approach zero as $r \to \infty$. Hence, for any $\delta > 0$ there is an r large enough so that $x^r \geq 0$ and

$$|F_i x^r| \leq \delta \qquad \text{for each} \quad i \ .$$

This completes the proof.

Observe that Corollary 6 shows a way of converging to an infimum of a convex and continuously differentiable program starting from a pair of primal-dual feasible vectors.

REFERENCES

[1]   Eaves, B. C.   "On the Basic Theorem of Complementarity,"
      Mathematical Programming 1, No. 1 (1971), 68-75.

[2]   Eaves, B. C., and Saigal, R.   "Homotopies for the Com-
      putation of Fixed Points on Unbounded Regions," Mathe-
      matical Programming 3, No. 2 (1972) 225-237.

[3]   Fisher, M. L., and Gould, F. J.   "A Simplicial Algo-
      rithm for the Nonlinear Complementarity Problem,"
      Mathematical Programming 6, No. 3 (1974) 281-300.

[4]   Garcia, C. B.   "A Fixed Point Theorem Including the
      Last Theorem of Poincare," to be published by Mathe-
      matical Programming.

[5]   Garcia, C. B., Lemke, C. E., and Leuthi, H. J.   "Sim-
      plicial Approximation of an Equilibrium Point of Non-
      Cooperative N-Person Games," Mathematical Programming,
      Eds. T. C. Hu and S. M. Robinson, Academic Press (1973)
      227-260.

[6]   Gould, F. J., and Tolle, J. W.   "Finite and Constructive
      Conditions for a Solution to f(x) = 0," University of
      Chicago, 1975.

[7]   Karamardian, S.   "The Complementarity Problem," Mathe-
      matical Programming 2, No. 1 (1972) 107-129.

[8]   Kuhn, H. W.   "A New Proof of the Fundamental Theorem of
      Algebra," Mathematical Programming Study 1, (1974)
      148-158.

[9]   Lemke, C. E.   "Bimatrix Equilibrium Points and Mathe-
      matical Programming," Management Science 11 (1965)
      681-689.

[10] Lemke, C. E., and Grotzinger, S. J.   "On Generalizing
      Shapley's Index Theory to Labelled Psuedomanifolds,"
      Rensselaer Polytechnic Institute, 1975.

[11] Lemke, C. E., and Howson, J. T., Jr. "Equilibrium
      Points of Bimatrix Games," SIAM Journal of Applied
      Mathematics 12, (1964) 413-423.

[12] Merrill, O. H. "Applications and Extensions of an Algo-
      rithm That Computes Fixed Points of Certain Non-Empty
      Convex Upper-Semi-Continuous Point-to-Set Mappings,"
      University of Michigan, 1971.

[13] Ortega, J. M., and Rheinboldt, W. C. Iterative Solu-
      tions of Nonlinear Equations in Several Variables,
      Academic Press, 1970.

[14] Scarf, H. E. (with the collaboration of T. Hansen)
      Computation of Economic Equilibria, Yale University
      Press, New Haven, 1973.

[15] Scarf, H. E. "The Approximation of Fixed Points of
      Continuous Mappings," SIAM Journal of Applied Mathe-
      matics 15 (1967) 1328-1343.

[16] Shapley, L. S. "A Note on the Lemke-Howson Algorithm,"
      Mathematical Programming Study 1 (1974).

A Fixed Point Approach to Stability in Cooperative Games

Alvin E. Roth

ABSTRACT

A generalization of the von Neumann-Morgenstern solution, called a <u>subsolution</u>, is defined. It is shown that a subsolution can be formulated as a fixed point with constraints. It is further shown that subsolutions exist for all cooperative games, and that for a general class of games which includes every game with a nonempty core, all subsolutions are nonempty.

1. <u>Introduction</u>

Game theory is the study of conflict situations involving several participants, or "players", with differing
objectives, each of whom has only partial control over the
outcome. Cooperative games are those in which the players
are free to communicate and reach binding agreements. Coalitions of players can thus form in an effort to further
some mutual interest.

No single model can adequately describe all cooperative games. In any model, however, we will need to develop
some criterion by which alternative outcomes can be compared.
In their discussion of how such a comparison ought to be
formulated, von Neumann and Morgenstern [9] write:

> This notion of superiority as between imputations
> ought to be formulated in a way which takes account
> of the physical and social structure of the milieu.
> That is, one should define that an imputation x
> is superior to an imputation y whenever this
> happens: Assume that society, i.e., the totality
> of all participants, has to consider the question
> whether or not to 'accept' a static settlement of
> all questions of distribution by the imputation
> y . Assume furthermore that at this moment the
> alternative settlement by the imputation x is
> also considered. Then this alternative x will
> suffice to exclude acceptance of y . By this
> we mean that a sufficient number of participants
> prefer in their own interest x to y , and are
> convinced of the possibility of obtaining the
> advantages of x . In this comparison of x to
> y the participants should not be influenced by
> the consideration of any third alternative (imputa-
> tions). I.e., we conceive the relationship of

superiority as an elementary one, correlating the
two imputations x and y only. The further
comparison of three or more -- ultimately of all --
imputations is the subject of the theory which must
now follow, as a superstructure erected upon the
elementary concept of superiority.

Whether the possibility of obtaining certain
advantages by relinquishing y for x , as discussed in the above definition, can be made con-
vincing to the interested parties will depend upon
the physical facts of the situation -- in the terminology of games, on the rules of the game.

We prefer to use, instead of 'superior' with its
manifold associations, a word more in the nature
of a terminus technicus. When the above described
relationship between two imputations x and y
exists, then we shall say that x dominates y .

In general, we shall say that an outcome x dominates an
outcome y if some coalition of players prefers x to y ,
and has sufficient power, according to the rules of the game,
to "move" the game from y "in the direction" of x .

1.1 Characteristic Function Form

To make these ideas more concrete, consider a simple
model of a game known as the characteristic function form,
with side payments. Such a game consists of a set N =
{1,...,n} of n players, and a real-valued function v
which is defined on the subsets (coalitions) of N . We interpret the function v as the worth of a coalition I $\subseteq$ N:
That is v(I) is the amount of utility that coalition I
can obtain for itself, regardless of the action of the players N - I . We assume that v($\emptyset$) = 0 . Implicit in our
definition of v is the assumption that the players in a coalition I are able to distribute any utility which they obtain freely among themselves. It is this assumption which we
refer to by the name of "side payments".

The set of outcomes, or __imputations__ of such a game is therefore the set

$$X = \{(x_1, \ldots, x_n) \mid \sum_{i \in N} x_i = v(N) \ , \quad x_i \geq v(\{i\}) \ , \quad \forall \ i \ \in \ N\} \ .$$

This represents the set of all ways in which the players can distribute the wealth available to them, in such a manner that every player does at least as well as he can do on his own.[1]

We say that for $x, y \in X$, x __dominates__ y __via a coalition__ $I \subseteq N$ if (i) $x_i > y_i$, $\forall \ i \ \in \ I$ and (ii) $\sum_{i \in I} x_i \leq v(I)$. Condition (i) asserts that every member of I prefers x to y , and condition (ii) asserts that I has the power to obtain the wealth it receives at the outcome x .

The modern theory of games can comfortably deal with models of considerably more generality than the one described above. For much of what follows, we will deal not with a detailed model, but merely with the abstract entities N, X and $\succ$.

1.2   Abstract Games

Consider an abstract game $(N, X, \succ)$, where N is the set of players, X is the set of possible outcomes, and $\succ$ is a binary relation on X called __domination__. The domination relation may be thought of as a force acting on the game, such that if for $x, y \in X$, $x \succ y$, the game has a tendency to "move" from the point y to the point x . We

---

[1] For simplicity we have made the implicit assumption that $v(N) \geq v(S) + v(N-S)$ for $S \subseteq N$.

are interested in finding sets of outcomes which are in some
sense stable with respect to domination. Our intuitive notion of stability leads us to look for sets on which no forces act, or on which all forces are in some sense "balanced".

We will find it convenient to define the following notation, relating to the domination relation on sets of outcomes: For every $x \in X$ define the dominion of x to be the
set $D(x) = \{y \in X | x \succ y\}$. For every $S \subseteq X$ define the domin-
ion of the set S to be $D(S) = \bigcup_{x \in S} D(x)$. Finally, for
each $S \subseteq X$, denote the complement of $D(S)$, the set of
outcomes undominated by S , by $U(S) = X - D(S)$.

1.3 The Core[1]

We can now define the core of a game to be the set of
outcomes undominated by any other outcome, i.e., the core is
the set $C = U(X)$. At a point in the core, no coalition of
players has both the power and the desire to force another
outcome. The core therefore possesses a very compelling kind
of stability -- once a point in the core is reached, there is
no tendency to move to another point. Nevertheless, the core
is not a sufficiently general concept to explain all of the
phenomena associated with cooperative games.

One of the factors which limits the usefulness of the
core concept is that it is easy to define games which have
an empty core. We will discuss one formulation which over-
comes this problem when we discuss the bargaining set.

A more fundamental criticism of the core, is that it
fails to take into account much of the cooperative, coalition forming behavior which cooperative games often exhibit.

---

[1]The concept of the core was introduced by Gillies
[2, 3] and Shapley [8].

For instance in economic games, outcomes in the core seem to
result when the players engage in "cut-throat" competition.
The theory of the core is therefore inadequate to explain the
formation of unions and cartels.

1.4 Solutions

A more comprehensive notion of stability was proposed
by von Neumann and Morgenstern in [9]. They defined a solution of a cooperative game to be a set $S \subseteq X$ such that:

(a) "No y contained in S is dominated by an x con-
     tained in S ".

(b) "Every y not contained in S is dominated by some x
     contained in S ".

Condition (a), which is called internal consistency, is
equivalent to the statement $S \subseteq U(S)$, while condition (b),
called external consistency, can be rewritten as $S \supseteq U(S)$.
Therefore we can equivalently define a solution to be a set
S such that:

(c) S = U(S) .

von Neumann and Morgenstern interpreted a solution as a
"standard of behavior" -- a rational form of behavior which,
once accepted by the players of the game, would prove to be
stable. They justified this, in part, as follows:

> Indeed, it appears that the sets of imputations S
> which we are considering correspond to the 'standard
> of behavior' connected with a social organization.
> Let us examine this assertaion more closely.

> Let the physical basis of a social economy be given,
> -- or, to take a broader view of the matter, of a
> society. According to all tradition and experience
> human beings have a characteristic way of adjusting
> themselves to such a background. This consists of
> not setting up one rigid system of apportionment,
> i.e., of imputation, but rather a variety of alter-
> natives, which will probably all express some
> general principles but nevertheless differ among

themselves in many particular respects. This
system of imputations describes the 'established
order of society' or 'accepted standard of behavior'.

Obviously no random grouping of imputations will
do as such a 'standard of behavior': It will have
to satisfy certain conditions which characterize
it as a possible order of things. This concept of
possibility must clearly provide for conditions of
stability. The reader will observe, no doubt, that
our procedure in the previous paragraphs is very
much in this spirit: The sets S of imputations
x,y,z ... correspond to what we now call 'standard
of behavior,' and the conditions (a) and (b), or
(c), which characterize the solution S express,
indeed, a stability in the above sense.

The disjunction into (a) and (b) is particularly
appropriate in this instance. Recall that domination of y by x means that the imputation x ,
if taken into consideration, excludes acceptance
of the imputation y (this without forecasting
what imputation will ultimately be accepted).
Thus, (a) expresses the fact that the standard of
behavior is free from inner contradictions: No
imputation y belonging to S -- i.e., conforming
with the 'accepted standard of behavior' -- can be
upset -- i.e., dominated -- by another imputation
x of the same kind. On the other hand (b) expresses
that the 'standard of behavior' can be used to dis-
credit any non-conforming procedure: Every imputation y not belonging to S can be upset -- i.e.,
dominated -- by an imputation x belonging to S .

Observe that we have not postulated that a y belonging to S should never be dominated by any
x . Of course, if this should happen, then x
would have to be outside of S , due to (a). In
the terminology of social organizations: An imputation x which conforms with the 'accepted standard
of behavior' may be upset by another imputation y ,
but in this case it is certain that y does not con-
form. It follows from our other requirements that
then y is upset in turn by a third imputation z

> which again conforms. Since x and z both con-
> form, z cannot upset x -- a further illustra-
> tion of the intransitivity of 'domination.'

> Thus our solutions S correspond to such 'stan-
> dards of behavior' as have an inner stability:
> Once they are generally accepted they overrule
> everything else and no part of them can be over-
> ruled within the limits of the accepted standards.
> This is clearly how things are in actual social
> organizations, and it emphasizes the perfect ap-
> propriateness of the circular character of our
> conditions in (a) and (b), or in (c).[1]

Thus a solution constitutes a set of outcomes which, once accepted, becomes self-enforcing. Outcomes outside of the solution are "overruled" because they are dominated by outcomes in the solution.

We have not yet addressed the question of existence -- whether or not a set S can be found which fills the re-quirements for a solution. von Neumann and Morgenstern con-jectured that a solution could be found for every game (in characteristic function form, with side payments). They at-tached great importance to this conjecture, going so far as to write:

> There can be, of course, no concessions as regards
> existence. If it should turn out that our require-
> ments concerning a solution S are, in any special
> case, unfulfillable, -- this would certainly neces-
> sitate a fundamental change in the theory. Thus a
> general proof of the existence of solutions S for
> all particular cases is most desirable.[2]

The effort to prove an existence theorem continued un-til 1968. In that year, Lucas [4] produced a counter-

---

[1]pp. 41-42. Text amended to remove typographical inconsistencies.

[2]p. 42.

example; a ten person game with side payments, which has a non-empty core, and for which no solution exists.

It is our intention, therefore, to introduce a generalization of the von Neumann-Morgenstern solution which preserves the interpretation of a standard of behavior; and for which existence can be proved.

2. Subsolutions

2.1 Interpretation

We define a subsolution of a game to be a set $S \subseteq X$ such that (i) $S \subseteq U(S)$; and (ii) $S = U^2(S)$.

The first condition, $S \subseteq U(S)$, says merely that the set S is internally consistent -- no element of S dominates any other element of S .

To understand the second condition, consider an arbitrary set $A \subseteq X$, and the set $U^2(A) = U(U(A))$. No point b in $U^2(A)$ is dominated by any point in the set $U(A)$. Therefore if b is dominated by some point x , then x is not contained in $U(A)$, but lies instead in its complement, $D(A)$. Therefore x is dominated by the set A : i.e., there is a point in A which dominates x .

We may therefore regard the set $U^2(A)$ as the set of points "protected" by A , in the sense that any point which dominates some point in $U^2(A)$ is in turn dominated by some point in A . We will say that a set $A \subseteq X$ is "self-protecting" if $A \subseteq U^2(A)$. Thus if $x \in A \subseteq U^2(A)$, and if $y \succ x$, then there exists $z \in A$ such that $z \succ y$.

Note that condition (ii) of the definition of a subsolution requires more than that S be self-protecting. It is not sufficient that $S \subseteq U^2(S)$, we require also that $S \supseteq U^2(S)$, i.e., that $S = U^2(S)$. It is this additional requirement which enables us to interpret a subsolution S as a "standard of behavior", in the sense of von Neumann and Morgenstern. In order to do this, we must show that once S is generally accepted, only outcomes in S can be considered "sound".

Suppose, then, that S is generally accepted. Any point in the set D(S) is "overruled" by a point in S , and hence unacceptable. The set of points not immediately overruled by S in this manner is U(S) . However, $S = U^2(S)$, and so every point in U(S) - S is dominated by some other point in U(S) - S . Therefore, such points overrule each other.

We see that S is stable with respect to the set U(S) in precisely the same way that the core, C = U(X) , is stable with respect to the entire set of outcomes X .

It is clear that the notion of a subsolution generalizes the von Neumann-Morgenstern solution, since if S is a solution, then $S = U(S) = U^2(S)$, and so S is also a subsolution. Furthermore, a solution S is clearly a maximal subsolution, since it is a maximal internally consistent set.

It is also clear that every subsolution contains the core, since points outside of S are either dominated by S or by U(S) - S , and hence cannot lie in the core.

In the next section we will prove a lattice fixed point theorem which, together with a corollary, demonstrates that (possibly empty) subsolutions exist for arbitrary abstract games. We shall also consider conditions which insure that nonempty subsolutions exist.

2.2 Lattices and Semilattices

A partially ordered set is a set P on which a binary
relation $\geq$ is defined such that $(\forall x,y,z \in P)$, (i) $x \geq$
x ; (ii) $x \geq y$ and $y \geq x \Rightarrow x = y$; and (iii) $x \geq y$ and
$y \geq z \Rightarrow x \geq z$.

A chain is a partially ordered set Q such that every
pair of elements is comparable: i.e., $(\forall x,y \in Q)$ either
$x \geq y$ or $y \geq x$.

A meet semilattice is a partially ordered set L such
that every two points $x,y \in L$ have a greatest lower bound
in L , denoted $x \wedge y$ and called their meet. A meet semilattice is said to be complete if every nonempty subset
$A \subseteq L$ has a greatest lower bound $\wedge\{A\}$ in L called its
meet.

A join semilattice is a partially ordered set L such
that every two points $x,y \in L$ have a least upper bound in
L denoted $x \vee y$ and called their join. A join semilattice
is said to be complete if every nonempty subset $A \subseteq L$ has a
least upper bound $\vee\{A\}$ in L called its join.

Note that if $x \leq y$ then $x \wedge y = x$ and $x \vee y = y$.

A lattice is a partially ordered set which is both a
meet and a join semilattice. It is said to be complete if
it is both a complete meet and a complete join semilattice.
It follows that a complete lattice has both a least element
and a greatest element.

A subset K of a lattice L is called a complete
meet subsemilattice if it is a complete meet semilattice and
the meet of any nonempty subset $S \subseteq K$ coincides with the
meet of the same set considered as a subset of L . A complete join subsemilattice is defined dually.

We shall also use the following variant of Zorn's
Lemma: "Every chain in a partially ordered set is contained

in a maximal chain."

2.3 **A Lattice Fixed-Point Theorem** (see [7]).

Let L be a complete lattice. Denote elements of L
with small letters $a,b,c,\ldots$, and denote subsets of L with
capital letters $A,B,C,\ldots$. Consider a function $U:L \to L$
with <u>property</u> P : For any $A \subseteq L$, $U(\vee A) = \wedge U(A)$, where
$U(A) \equiv \{U(a)\,|\,a \in A\}$. Denote the composition of U with
itself by U^2 .

<u>Lemma 2.1</u>: (1) $a \leq b \Rightarrow U(a) \geq U(b)$; (2) $a \leq b \Rightarrow U^2(a) \leq$
$$U^2(b) .$$

<u>Proof</u>: (1) $a \leq b \Rightarrow b = a \vee b \Rightarrow U(b) = U(a) \wedge U(b) \leq U(a)$.

(2) Apply part (1) to $U(a)$ and $U(b)$.

Define $L_D(U) \equiv \{a \in L\,|\,a \leq U(a)\}$ and $L_D(U^2) \equiv$
$$\{a \in L\,|\,a \leq U^2(a)\} .$$

<u>Lemma 2.2</u>: (1) $U^2:L_D(U) \to L_D(U)$; (2) $U^2:L_D(U^2) \to L_D(U^2)$.

<u>Proof</u>: (1) $a \in L_D(U) \Rightarrow a \leq U(a) \Rightarrow U(a) \geq U^2(a) \Rightarrow U^2(a) \leq$

$$U(U^2(a)) \Rightarrow U^2(a) \in L_D(U) .$$

(2) $a \in L_D(U^2) \Rightarrow a \leq U^2(a) \Rightarrow U^2(a) \leq U^2(U^2(a)) \Rightarrow$

$$U^2(a) \in L_D(U^2) .$$

<u>Lemma 2.3</u>: $L_D(U^2)$ is a complete join subsemilattice of L .

<u>Proof</u>: We need to show that for any subset $A \subseteq L_D(U^2)$,

$\vee\{A\} \in L_D(U^2)$. We show this as follows:

$A \subseteq L_D(U^2) \Rightarrow [\vee a \in A]$, $a \leq U^2(a) \Rightarrow \vee\{A\} \leq$

$\vee\{U^2(A)\}$. Now $a \leq \vee\{A\} \vee a \in A \Rightarrow U^2(a) \leq$

$$U^2(\vee\{A\}) \vee a \in A \implies \vee\{U^2(A)\} \leq U^2(\vee\{A\}) \quad . \quad So$$

$$\vee\{A\} \leq U^2(\vee\{A\}) \implies \vee\{A\} \in L_D(U^2) \quad .$$

In a similar manner, it can be shown that $L_D(U)$ is a complete meet subsemilattice of L, but this is not needed for our proof.

We know from Tarski's theorem (see Birkhoff [1]) that U^2 has a fixed point in L, while it is not generally true that U has a fixed point. We can, however, state the following:

Theorem 2.1: There exists an element $s \in L$ such that $s = U^2(s)$ and $s \leq U(s)$.

Proof: Let $B \equiv L_D(U) \cap L_D(U^2)$. B is nonempty, since inf $L \in B$. We need to show that there exists an $s \in B$ such that $s = U^2(s)$.

Let M be a maximal chain in B , and let $s = \vee\{M\}$. Then $s \in L_D(U^2)$ since $M \subseteq L_D(U^2)$ and $L_D(U^2)$ is a complete join subsemilattice. In order to show that $s \in B$, we must demonstrate that $s \in L_D(U)$.

Suppose not. Then $s \nleq U(s)$. Recall that $s = \vee\{M\}$, so $U(s) = \wedge\{U(M)\}$, and $s \nleq U(s) \implies \exists m \in M \ni s \nleq U(m)$. So $U(m)$ is not an upper bound to $s = \vee\{M\}$, and consequently there exists an element $n \in M \ni n \nleq U(m)$.

Now $M \subseteq L_D(U) \implies m \leq U(m)$ and $n \leq U(n)$, and M is a chain, so either $m \geq n$ or $n \geq m$.

But $n \nleq m$, since $n \nleq U(m)$. And if $m \leq n$, then by Lemma 2.1, $U(m) \geq U(n) \geq n$, which contradicts the fact that $n \nleq U(m)$.

Therefore, $s \leq U(s)$, which is to say $s \in L_D(U)$.

To complete the proof, we demonstrate that $s = U^2(s)$. By Lemma 2.2, $U^2(s) \in B$ · $s \in L_D(U^2) \Rightarrow s \leq U^2(s)$. If $s < U^2(s)$, then the chain M is not maximal, contrary to the assumption.

Examination of the proof reveals that the element $s \in L$ produced above is a maximal fixed point of the required kind: i.e., if $t \in L$ and $t = U^2(t) \leq U(t)$, then $t \not\geq s$. A sufficient condition to insure $s \neq \inf L$ is that there exists an $m \in B$ distinct from $\inf L$.

2.4 Existence of Subsolutions

To see that this theorem applies to our formulation of a subsolution, let us consider, for arbitrary X and $\succ$, the function $U = X-D$ defined previously. For each $S \subseteq X$, $U(S) = X - D(S)$, where $D(S) = \bigcup_{x \in S} D(x)$. Let $L = 2^X$ be the complete lattice of subsets of X , ordered by set inclusion. Then meet and join on L are ordinary set intersection and union, and $\inf L = \emptyset$, the empty set.

If $A \subseteq L$, then A is a collection of subsets $S_a \subseteq X$. So

$$D(\vee A) = D\left(\bigcup_{S_a \in A} S_a\right) = \bigcup_{x \in \cup S_a} D(x) = \bigcup_{S_a \in A} [\bigcup_{x \in S_a} D(x)] =$$

$\cup D(S_a)$, and therefore

$$U(\vee A) = \cap U(S_a) = \wedge U(A) \quad .$$

Thus U has property P , and we can state the following:

Corollary 2.1: There exists a subset $S \subseteq X$ such that

$$S = U^2(S) \quad \text{and} \quad S \subseteq U(S) \quad .$$

Furthermore, a sufficient condition to insure that the set
S is nonempty is the existence of a non-empty set $T \subseteq X$
such that $T \subseteq U(T)$ and $T \subseteq U^2(T)$.

Examining the lattice formulation of this section in
the context of game theory, we see that the set $L_D(U)$ is
precisely the set of internally consistent subsets of X .
The fact that it is a meet subsemilattice of L is now
merely a result of the observation that intersections of
internally consistent sets are internally consistent.

A key part of the proof involved showing that the set
$s = \cup\{m \in M\}$ is internally consistent. In a game theoretic
context, the argument can be stated as follows: If the set
$\cup\{m \in M\}$ is not internally consistent, it contains elements
x and y such that $y \in D(x)$. Since M is a chain, there
exists a set $m \in M$ such that the elements $x, y \in m$. This
contradicts the internal consistency of m . Hence the set
$\cup\{m \in M\}$ is internally consistent.

REFERENCES

[1] Birkhoff, G., Lattice Theory, (3rd ed.), American Mathe-
     matical Society Colloquium Publications, Vol. 25, 1967.

[2] Gillies, D. B., "Locations of Solutions", in "Report of
     an Informal Conference on the Theory of N-Person Games",
     edited by H. W. Kuhn, Princeton University, March, 1953.

[3] Gillies, D. B., "Some Theorems on N-Person Games", Ph.D.
     Thesis, Department of Mathematics, Princeton University,
     1953.

[4] Lucas, W. F., "A Game With No Solution", Bulletin of
     the American Mathematical Society, 74 (1968), pp. 237-
     239.

[5] Roth, A. E., "Subsolutions of Cooperative Games", Tech-
     nical Report No. 118, Institute for Mathematical
     Studies in the Social Sciences, Stanford University
     (also Technical Report 73-12, Department of Operations
     Research), December, 1973.

[6] Roth, A. E., "Topics in Cooperative Game Theory", Tech-
     nical Report SOL 74-8, Systems Optimization Laboratory,
     Department of Operations Research, Stanford University,
     July, 1974.

[7] Roth, A. E., "A Lattice Fixed-Point Theorem with Con-
     straints", Bulletin of the American Mathematical Soci-
     ety, Vol. 80, No. 6, November 1974.

[8] Shapley, L. S., "Open Questions", in "Report of an In-
     formal Conference on the Theory of N-Person Games",
     edited by H. W. Kuhn, Princeton University, March, 1953.

[9] von Neumann, J., and Morgenstern, O., Theory of Games
     and Economic Behavior, Princeton, (1944).

Error Bounds for Approximate Fixed Points

Christopher Bowman
and
Stepan Karamardian

ABSTRACT

   The error bound between a given function and its approximate fixed-point, as computed through finite pivoting algorithms, is usually expressed in terms of max-norm. This is because the manner in which the integer labeling is defined lends itself to a simple and direct derivation of such a max-norm error bound. However, such norms are not invariant under rotation of coordinate axes. In this paper the best uniform error bound, expressed in euclidean norm, for a proper integer labeling of a triangulation of the standard n-simplex S^n is derived.

1. Introduction

The Brouwer fixed point theorem, first proved in 1912, states that a continuous function f from a simplex into it- self has at least one fixed point. Since then, considerable effort has been directed towards developing an efficient algorithm for computing an approximate fixed point. In 1967, H. Scarf [4] presented such an algorithm. This algorithm utilizes a systematic pivoting procedure on a class of sub- simplices of the given simplex, called "primitive" sets. In the following year, H. Kuhn [3] proposed a similar algorithm, where the pivoting is carried on a simplicial subdivision (triangulation) of the given simplex.

In H. Kuhn's algorithm, for example, the regular n- simplex S^n is triangulated with a given mesh. Each vertex of each n-simplex in the triangulation is "properly" labelled with an integer 0 through n . The algorithm terminates whenever the pivoting procedures reaches a "completely" la- belled n-simplex, i.e., a simplex whose n + 1 vertices are distinctly labelled. After applying an appropriate initia- tion procedure one obtains a starting (n-1)-simplex, To^{n-1} which is a face of an initial n-simplex To^n . The n-vertices of To^{n-1} are distinctly labelled. If To^n is also distinctly labelled, the algorithm terminates. Other- wise that vertex of To^{n-1} which carries the same label as the vertex of To^n which was adjoined to To^{n-1} to form

To^n , is dropped. The remaining n vertices of To^n determine a unique (n-1)-simplex T_1^{n-1} which is a face of a
unique n-simplex T_1^n adjacent to To^n . The same procedure
is continued till a completely labeled n-simplex T_k^n is
reached. It can be shown, that this algorithm terminates in
a finite number of iterations.

Each point x in a completely labeled n-simplex is
characterized as an approximate fixed point to f . The
"difference" between x and f(x) depends on the mesh of the
triangulation and on the type of labeling chosen to label the
traingulation. The bound on this difference is usually
expressed in terms of the max-norm, due to the coordinate
nature of the integer labeling. Such bounds, however, are
not invariant under coordinate rotation. In this paper we
present the best uniform error bound on $\| f(x)-x \|$ where
$\| \cdot \|$ denotes the euclidean norm.

1. Definitions and Notations

Let $v^o, v^1, \ldots, v^n$ be n + 1 linearly independent vectors in R^{n+1}. The set

$$T^n = \left\{ x \varepsilon R^{n+1} \middle| x = \sum_{i=0}^{n} \lambda_i \, v^i, \ \sum_{i=0}^{n} \lambda_i = 1, \ \lambda_i > 0, \ i=0,\ldots,n \right\}$$

is called an n-simplex generated by $v^o, v^1, \ldots, v^n$. The
vectors $v^o, v^1, \ldots, v^n$ are called the vertices of T^n. An
n-simplex is called <u>regular</u> if the euclidean distance between
every pair of its vertices is constant. An example of a

regular n-simplex is the <u>standard</u> n-simplex S^n generated by the unit vectors $e^0, e^1, \ldots, e^n$ in R^{n+1} .

The r-simplex T^r , $0 \leq r \leq n$, generated by the vectors $v^{i_0}, \ldots, v^{i_r}$, where $\{i_0, \ldots, i_r\} \subset \{0, 1, \ldots, n\}$, is called an <u>r-face</u> of T^n . A pair of disjoint faces of T^n are called <u>opposite faces</u> if every vertex of T^n is a vertex of one, and only one, of the two faces.

A <u>triangulation</u> K^n of an n-simplex T^n is a partitioning of the closure $\overline{T^n}$ of T^n by a finite set of mutually disjoint simplices such that

1. every face of every simplex in K^n is an element of K^n , and

2. the point set union of the elements of K^n is the closure of T^n .

Let K^n be a triangulation of the standard n-simplex S^n , and let K^0 denote the set of all 0-simplices in K^n . An <u>integer labeling</u> of K^n is a function from K^0 into the set $\{0, 1, \ldots, n\}$. An integer labeling ℓ of K^n is called <u>proper</u> if for every $v \in K^0$ we have

$$\ell(v) = i \implies v_i > 0 \quad , \quad i = 0, 1, \ldots n$$

where v_i is the ith component of v and i is called the label of v .

An   n-simplex T^n in K^n is called   <u>complete</u> with respect to a given proper integer labeling if its vertices are
distinctly labelled, i.e., no two vertices of T^n receive
the same label under ℓ .

The   <u>mesh</u> (or diameter) of a set of points is the supremum of the distances between every pair of points in the set.
The mesh of a triangulation K^n of S^n is the largest mesh
of the simplices of K^n .

2.   Error Bounds for Approximate Fixed Points

Given a function,   f   from $\overline{S}^n$ into itself and an
$\alpha \geq 0$ then an $x \in \overline{S}^n$ satisfying $\|f(x)-x\| \leq \alpha$ is called an α-approximate <u>fixed point</u> of   f   .   And if $\alpha = 0$
then   x   is a <u>fixed point</u> of   f   .

The main result of this paper is the following:

<u>Theorem</u>:   Given a continuous function, $f:\overline{S}^n \to \overline{S}^n$, and the
proper integer labeling

$$\ell(v) = \min \{i \mid v_i > 0 \text{ and } f_i(v) \leq v_i\}$$

of a triangulation K^n of S^n with mesh δ , let T^n
be a complete   n-simplex in K^n satisfying $\|f(x)-f(y)\| \leq \varepsilon$
for every   x   and   y   in T^n .   Then for any point   x   in
T^n

$$\|f(x)-x\| \leq \sqrt{n(n+1)/2} \ (\delta+\varepsilon) \quad .$$

The above error bound is achieved for some f and K^n .

The following three lemmas will be used in the proof of the theorem.

Lemma 1: (Sperner, 1928). Given any proper integer labeling of a triangulation K^n of S^n , there exists at least one complete n-simplex in K^n .
Proof: See Cohen [2].

Lemma 2: (Jung, 1901). Any set of points of mesh d in an n-dimensional space can be enclosed in an n-dimensional sphere of radius $\sqrt{n/2(n+1)}$ d . And this is the radius of the circumscribed n-sphere of a regular n-simplex of mesh d .
Proof: See Blumenthal and Wahlin [1].

Lemma 3: The mesh of a regular n-simplex with an inscribed n-sphere of radius r is $\sqrt{2n(n+1)}$ r . The distance from the vertex of a regular n-simplex of mesh d to the centroid of its opposite face is $\sqrt{(n+1)/2n}$ d .
Proof: These results follow directly by assigning coordinates to the appropriate points in a regular n-simplex.

Proof of Theorem: The existence of a complete n-simplex T^n is guaranteed by Lemma 1. (In fact a pivoting algorithm for finding a complete n-simplex is exhibited in Kuhn [3].)

Define $g(x) = f(x)-x$, then for any x and y in T^n the following holds

$$\|g(x)-g(y)\| \leq \|f(x)-f(y)\| + \|x-y\| \leq \epsilon+\delta \quad .$$

By Lemma 1 the image of T^n under g is contained in a closed n-sphere $\sum^n$ with radius $\sqrt{n/2(n+1)}(\delta+\epsilon)$, where $\sum^n$ lies in the subspace $H^n = \left\{ x \mid x\epsilon R^{n+1} \text{ and } \sum_{i=0}^{n} x_i = 0 \right\}$. For

each i define $m_i = \min_{x\epsilon\Sigma^n} x_i$ and let $\Delta^n = \bigcap_{i=0}^{n} \left\{ x \mid x\epsilon R^{n+1} , \right.$

$\sum_{i=0}^{n} x_i = 0$, and $\left. x_i \geq m_i \right\}$. For each i-labeled vertex

v^i of T^n the given labeling yields

$$g(v^i)\epsilon\left\{ x \mid x\epsilon R^{n+1} , \sum_{k=0}^{n} x_k=0 , x_i\leq 0 , \text{ and } x_j\geq 0 , \text{ for all } j<i \right\} .$$

Now $m_i \leq g(v^i) \leq 0$ for all i , consequently the origin

is in Δ^n . If $g(v^n) = 0$, then the proof of the theorem

is clear. On the other hand if $g(v^n) \neq 0$, then Δ^n is

the translation into H^n of the closure of a regular n-sim-

plex. To prove this it suffices to show that Δ^n is the

convex hull of

$$\left\{ u^i \mid u^i = m - Me^i \quad \text{where} \quad m = \sum_{i=0}^{n} m_i e^i \quad \text{and} \right.$$

$$\left. M = \sum_{i=0}^{n} m_i \quad \text{for} \quad i = 0,1,\ldots,n \right\} \subset H^n .$$

and that $\|u^i - u^j\|$ is a nonzero constant, for all $i \neq j$.

Since $\sum_{i=0}^{n} g_i(v^n) = 0$ and $g_i(v^n) \geq 0$ for all i

less than n , the assumption that $g(v^n) \neq 0$ implies that

$g_n(v^n) < 0$. Consequently $M \leq m_n < 0$, and $\|u^i - u^j\| = $

$\sqrt{2M^2} > 0$ for all i and j . The following will show that

Δ^n is the convex hull of $\{u^i \mid i = 0,.,\ldots,n\}$. For any

point $x \in \Delta^n$, $x = \sum_{i=0}^{n} \lambda_i e^i$ where $\sum \lambda_i = 0$ and

$e^i = (M-u^i)/M$. Using $\sum_{i=0}^{n} m_i u^i = 0$ one obtains

$$x = \sum_{i=0}^{n} \lambda_i (m-u^i)/M = -\sum_{i=0}^{n} \lambda_i u^i/M + \sum_{i=0}^{n} m_i u^i/M = \sum_{i=0}^{n} ((m_i - \lambda_i)/m)u^i$$

Since $(m_i - \lambda_i)/M \geq 0$ and $\sum_{i=0}^{n} (m_i - \lambda_i)/M = 1$, x is a convex

combination of $\{u^i \mid i=0,1,\ldots,n\}$. Next let x be in the

convex hull of $\{u^i \mid i=0,1,\ldots,n\}$, then

$$x = \sum_{i=0}^{n} \gamma_i (m-Me^i) = m - M \sum_{i=0}^{n} \gamma_i e^i \geq m \quad \text{and} \quad \sum_{i=0}^{n} x_i = \sum_{i=0}^{n} m_i$$

$-M\sum_{i=0}^{n} \gamma_i = 0$. So $x \in \Delta^n$. Therefore Δ^n is the transla-

tion of the closure of a regular n-simplex.

$\sum^n$ is the inscribed n-sphere of Δ^n , since

$\sum^n \subseteq \Delta^n$ and for each i there exists an s ϵ $\sum^n$ such

that $s_i = m_i$. So,

$$\max \left\{ ||x-s|| \mid x \epsilon \Delta^n \text{ and } s \epsilon \sum^n \right\}$$

is attained when x is a vertex of Δ^n and s is the point

of tangency of $\sum^n$ at the centroid of the face of Δ^n op-

posite x . By Lemma 2 this distance is $\sqrt{n(n+1)/2}$ $(\delta+\epsilon)$.

Since $0 \epsilon \Delta^n$ the following holds,

$$||f(x)-x|| = ||g(x)-0|| \leq \max \left\{ ||x-s|| \mid x\epsilon\Delta^n \ s\epsilon\sum^n \right\} = \sqrt{n(n+1)/2} \ (\delta+\epsilon)$$

for any x ϵ T^n .

The following example will show that no smaller bound

will suffice with any δ and ϵ such that $(n/\sqrt{2})(\delta+\epsilon) \leq 1$.

Let T^n be the regular δ-mesh n-simplex with vertices

$\{v^0, v^1, \ldots, v^n\}$ such that $v^i = (\delta/\sqrt{2})e^i - (1-\delta/\sqrt{2})e^n$.

Define a function f as follows,

$$f(v^i)-v^i = \left[\sum_{\substack{j=0 \\ j \neq i}}^{n} (1/\sqrt{2})(\delta+\epsilon)e^j \right] - (n/\sqrt{2})(\delta+\epsilon)e^n \qquad i=0,1,\ldots,n$$

and linearly extend f such that f is a continuous map of

$\overline{S}^n$ into itself.

Clearly v^i is labeled i, so T^n is complete. And by the definition of T^n and f the following holds for all i and j .

$$\| f(v^i) - f(v^j) \|^2 = (v_i^{\,i} - (\delta+\epsilon)/\sqrt{2}\,)^2 + (v_j^{\,j} - (\delta+\epsilon)/\sqrt{2}\,)^2 = \epsilon^2 .$$

Thus T^n and f satisfy the hypotheses and

$$\| f(v^n) - v^n \| = \sqrt{n(n+1)/2}\ (\delta+\epsilon) .$$

So the given error bound cannot be improved. This completes the proof.

REFERENCES

[1] Blumenthal, L. M., and Wahlin, G. E. "On the Spherical Surface of Smallest Radius Enclosing a Bounded Subset of N-Dimensional Euclidean Space", Bull. Amer. Math. Soc., Vol. 47 (1941), p. 771-777.

[2] Cohen, D. I. A., "On the Sperner Lemma", Journal Combinatorial Th., 2, 1967, p. 585-587.

[3] Kuhn, H. W., "Simplicial Approximation of Fixed Points", Proc. Nat. Acad. Sci., USA, 61, 1968, p. 1238-1242.

[4] Scarf, H. "The Approximation of Fixed Points of a Continuous Mapping", SIAM J. Applied Math., 15, 1967, p. 1328-1343.

Computation of Fixed Point in a Nonconvex Region

Hisakazu Nishino

ABSTRACT

A nonconvex polyhedron X is defined as the union of compact convex polyhedra $X_1, X_2, \ldots, X_m$ in R^n . The present note gives a simple homeomorphism between X and some X_i which makes possible the use of a Brouwer or Kakutani fixed point algorithm in order to compute a fixed point of a continuous function $f: X \to X$ if the X_i's satisfy certain combinatorial conditions.

0. Notation

Let R^n denote the Euclidean n-space. A vector x in R^n is assumed to be a column but whenever there is no ambiguity the transpose sign is omitted. The usual inner product of x and y in R^n is denoted by xy . The interior of a set $Y \subset R^n$ is denoted by int Y . We will consider m nonempty compact convex polyhedra $X_1, X_2, \ldots, X_m$ such that

$$X_i = \left\{ x \in R^n \,\middle|\, A^i x \leq b^i \right\} \quad , \qquad i = 1, 2, \ldots, m \quad ,$$

where A^i denotes a $p_i \times n$ matrix and $b^i \in R^{p_i}$. Let $I = \{1, 2, \ldots, m\}$. We define

$$X_J = \bigcap_{i \in J} X_i \quad \text{and}$$

$$X^J = \bigcup_{i \in J} X_i \quad , \text{ where } J \subset I \quad .$$

We shall abbreviate X^I by X . Finally by A_k , we denote the k-th row of a matrix A .

1. Introduction

Let X be the union of nonempty convex polyhedra $X_1, X_2, \ldots, X_m$ in R^n . We shall impose on the X_i's the following conditions:

<u>Condition 1.</u> For any $J \subset I$, $X_J \neq \emptyset$ implies int $X_J \neq \emptyset$.

<u>Condition 2.</u> For any i and j in I , there is a sequence $(k_1, k_2, \ldots, k_p)$ of distinct indices such that $k_1 = i$, $k_p = j$ and $X_{\{k_s, k_{s+1}\}} \neq \emptyset$, $s = 1, 2, \ldots, p-1$.

<u>Condition 3.</u> For any $J \subset I$, there is an $i \in J$ such that it belongs to exactly one maximal star-shaped segment S of J where by a maximal star-shaped segment S of J we mean a subset of J such that $X_S \neq \emptyset$ and $X_T = \emptyset$ for every $T \subset J$ which includes S as a proper subset.

Let $f: X \to X$ be continuous. The problem which we shall consider in this paper is that of computing a fixed point $x^* = f(x^*) \in X$ of the function f under Conditions 1 to 3. This is a special case of the Eilenberg and Montgomery's scheme in [4].

It follows immediately from Conditions 1 to 3 that each X_i is homeomorphic to X . Suppose for the time being a homeomorphism $h: X \to X_i$ is obtained for some $i \in I$, and define $g: X_i \to X_i$ by $g(y) = h(f(h^{-1}(y)))$, $y \in X_i$. The bicontinuity of h ensures that g is continuous on X_i and, therefore, has a Brouwer fixed point $y^* = g(y^*) \in X_i$. Since h is one-to-one, $y^* \in X_i$ is a fixed point of g if and only if $x^* = h^{-1}(y^*) \in X$ is a fixed point of f . It will be also obvious that if $y' \in X_i$ is an approximate fixed point of g then $x' = h^{-1}(y')$ gives the same order of approximation of a fixed point of f .

We will give in what follows a simple iterative method of constructing the homeomorphism h between X and some

X_i . An algorithm for calculating the value of g at every desired point of X_i is then shown, which we shall call Algorithm I. By Algorithm II, we shall denote the part of Algorithm I which determines the inverse image of h . Applying Algorithm I to a Brouwer or Kakutani fixed point algorithm established by Scarf [7], Kuhn [6], Hansen and Scarf [5], and Eaves [1], [2], and [3], we obtain an approximation of $y^* = g(y^*) \in X_i$. Algorithm II transforms it into the approximate value of our fixed point $x^* = f(x^*) \in X$.

2.   Reduction of Indices

The construction of our homeomorphism is based upon the following reduction process of indices $i \in I$:

(1)   Let   r = 1   and set   T(1) = I .

(2)   Find an $i(r) \in T(r)$ such that $X_{P(r)} \neq \emptyset$, where
$P(r) = \{j \in T(r) \mid X_{\{i(r),j\}} \neq \emptyset\}$.   Clearly, P(r)

which corresponds to   i(r)   is a maximal star-shaped segment of   T(r)   including   i(r) .   Stop the procedure if $P(r) = \{i(r)\}$.   Otherwise go to (3).

(3)   Put $T(r+1) = T(r) - \{i(r)\}$, replace   r   by   r+1 and return to (2).

It is easy to see that the above (1) to (3) reduce the index set   I   to a one point set $\{i(m)\}$ if and only if Conditions 2 and 3 hold.   It should be noted that each $i(r) \in T(r)$ can be found by the repeated use of Phase I of the simplex method.

We shall assign a point $x(r) \in int\ X_{P(r)}$ to each   i(r) , r = 1,2,...,m-1 , which is possible by Condition 1.   Since

Phase I calculates an extreme point of $X_{P(r)}$ whenever it is
nonempty, this assignment can be done by perturbing the extreme point towards int $X_{P(r)}$ in the usual way. In case of
$P(r) \subset P(r')$, $r > r'$, it will be convenient to put
$x(r) = x(r')$.

The following is an example of the algorithm which
yields the sequences $\{(i(1), i(2), \ldots, i(m)); (P(1), P(2), \ldots, P(m-1)); (x(1), x(2), \ldots, x(m-1))\}$:

   i.   Let $r = 1$ and set $T(1) = I$.

  ii.   Let $p = 1$ and set $U(r,1) = T(r)$.

 iii.   Let $j(r,p)$ be the smallest index in $U(r,p)$.
Obtain the set $D(r,p) = \{i \epsilon T(r) \mid$ there exists a
$z \epsilon R^n$ such that $A^{j(r,p)} z \le b^{j(r,p)}$ and $A^i z \le b^i\}$.
If $D(r,p) \subset P(r')$ for some $r' < r$, put
$P(r) = D(r,p)$, $x(r) = x(r')$ and go to viii.
Otherwise go to iv.

  iv.   Find an extreme point $a(r,p)$ of the convex polyhedron $V(r,p) = \{z \epsilon R^n \mid A^i z \le b^i$ for every
$i \epsilon D(r,p)\}$. If $V(r,p) = \emptyset$, go to v. Otherwise
put $a(r,p) = z(r)$, $i(r) = j(r,p)$ and $P(r) = D(r,p)$, and go to vi.

   v.   Put $U(r,p+1) = U(r,p) - \{j(r,p)\}$. Stop the procedure if $U(r,p+1) = \emptyset$ (Condition 3 is violated).
Otherwise increase p and return to iii.

  vi.   Solve the equation $A^i_k u = -1$, $(i,k) \epsilon B(r)$, where
$B(r) = \{(i,k) \mid i \epsilon P(r), 1 \le k \le n$, and $A^i_k z(r) = b^i_k\}$.
Stop the procedure if the equation has no solution
(Condition 1 is violated). Otherwise denote the
solution by $u(r)$ and go to vii.

vii.   If $r = m$, go to ix. Otherwise calculate the
positive number $\varepsilon(r)$ such that

$$\varepsilon(r) = \min_{i \in P(r)} \ \min_{k \in K(i)} \ \frac{b_k^i - A_k^i z(r)}{A_k^i u(r)}$$

where $K(i) = \{k \,|\, A_k^i u(r) > 0\}$,

and put $x(r) = z(r) + \frac{1}{2}\varepsilon(r)u(r)$, and go to viii.

viii. If $P(r) - \{i(r)\} = \emptyset$ and $r < m$, stop the procedure (Condition 2 is violated). If $r = m$ go to
ix. Otherwise put $T(r+1) = T(r) - \{i(r)\}$, increase r , and return to ii.

ix.   Take out the sequences $\{(i(1),i(2),\ldots,i(m))$;
$P(1),P(2),\ldots,P(m-1))$; $(x(1),x(2),\ldots,x(m-1))\}$.
Stop the procedure.

3. Homeomorphism

Let us define functions $h_r : X^{T(r)} \rightarrow R^n$, $r = 1,2,\ldots,m-1$
as

$$h_r(x) = \begin{cases} x, & \text{if } x \in X^{T(r)} - X_{i(r)} \\ x(r), & \text{if } x = x(r) \\ x(r) + \dfrac{\lambda(r,x)}{\mu(r,x)}(x-x(r)), & \text{if } x \in X_i(r) \text{ and } x \neq x(r) \end{cases}$$

where $\lambda(r,x)$ is the maximum value of $\lambda > 0$ such that
$x(r) + \lambda \cdot (x-x(r)) \in X^{P(r) - \{i(r)\}}$ and $\mu(r,x)$ the maximum

value of $\mu > 0$ such that $x(r) + \mu \cdot (x-x(r)) \epsilon X^{P(r)}$. It is

clear that $\mu(r,x) \geq 1$ and $x(r) + \lambda(r,x)(x-x(r)) \epsilon X^{P(r)-\{i(r)\}}$

for every $x \epsilon X_{i(r)}$ such that $x \neq x(r)$. Since $X^{P(r)}$ is

star-shaped at $x(r)$, we see $h_r(x) \epsilon X^{T(r+1)}$. Note that

$\frac{\lambda(r,x)}{\mu(r,x)} \leq 1$ for any $x \neq x(r)$ in $X_{i(r)}$. Letting

$\nu(r,x) = \frac{\lambda(r,x)}{\mu(r,x)}$, we obtain, for every $x \neq x(r)$ in $X_{i(r)}$,

$$h_r(x) = \nu(r,x)x + (1-\nu(r,x))x(r), \quad 0 < \nu(r,x) \leq 1 \quad .$$

This implies that h_r is continuous at $x(r)$. We have,

furthermore, $\nu(r,x) = 1$ for every x in $X_{i(r)} \cap cl(X^{T(r)}$

$- X_{i(r)})$, where $cl(Y)$ stands for the closure of a set

$Y \subset R^n$. Since $\lambda(r,x)$ and $\mu(r,x)$ are both continuous on

$X_{i(r)} - \{x(r)\}$, we ultimately have the continuity of h_r

on $X^{T(r)}$. Now, it is easy to see that h_r is a homeomor-

phism between $X^{T(r)}$ and $X^{T(r+1)}$. In fact, a continuous

function $d_r : X^{T(r+1)} \to X^{T(r)}$ defined by the following gives

the inverse of h_r :

$$d_r(y) = \begin{cases} y \ , & \text{if } y \epsilon X^{T(r+1)} - X_{i(r)} \\ x(r) & \text{if } y = x(r) \\ x(r) + \dfrac{1}{(r,y)}(y-x(r)), & \text{if } y \epsilon X_{i(r)} \cap X^{T(r+1)} \text{ and} \\ & \quad y \neq x(r) \ . \end{cases}$$

Let h be the composite mapping $h_1 h_2 \ldots h_{m-1}: X \to X_{i(m)}$.
Then, h is the desired homeomorphism.

4. Algorithm

We shall now describe the algorithm which calculates
the value of $g(y)$ at any desired point y in $X_{i(m)}$.
Assume the sequences $\{(i(1),i(2),\ldots,i(m)); (P(1),P(2),\ldots,$
$P(m-1)); (x(1),x(2),\ldots,x(m-1))\}$ to be given.

Algorithm I:

    i.   Pick a point $y \epsilon X_{i(m)}$.

   ii.   Let $q = m-1$ and put $y(m-1) = y$.

  iii.   If $y(q)$ satisfies $A^{i(q)} y(q) \leq b^{i(q)}$, go to iv.
        Otherwise put $y(q-1) = y(q)$ and go to v.

   iv.   If $y(q) = x(q)$, put $y(q-1) = y(q)$. If
        $y(q) \neq x(q)$, compute the following positive
        numbers:

$$\lambda(q,y(q)) = \underset{i \epsilon P(q) - \{i(q)\}}{\max} \underset{k \epsilon K(i)}{\min} \frac{b_k^i - A_k^i x(q)}{A_k^i(y(q)-x(q))}$$

$$\mu(q,y(q)) = \underset{i \epsilon P(q)}{\max} \underset{k \epsilon K(i)}{\min} \frac{b_k^i - A_k^i x(q)}{A_k^i(y(q)-x(q))}$$

where $K(i) = \{k | A_k^i(y-z(q)) > 0\}$, and put $y(q-1) = x(q)$
$+ \dfrac{1}{\nu(q,y(q))} (y(q) - x(q))$.

v.   If $q > 1$, replace q by q-1 and return to
      iii. If $q = 1$, go to vi.

vi.   Let $r = 1$, put $w(1) = f(y(0))$.

vii. If $w(r)$ satisfies $A^{i(r)}w(r) \leq b^{i(r)}$, go to
      viii. Otherwise put $w(r+1) = w(r)$ and go to ix.

viii. If $w(r) = x(r)$, put $w(r+1) = w(r)$. If
      $w(r) \neq x(r)$, compute $\lambda(r,w(r))$ and $\mu(r,w(r))$
      in the same way as in iv, and put $w(r+1) =$
      $x(r) + \nu(r,w(r))(w(r) - x(r))$.

ix.   If $r < m-1$, replace r by r+1 and return to
      vii. If $r = m-1$, go to x.

x.   Put $g(y) = w(m)$. Stop the procedure.

Note that steps i to v above constitute Algorithm II for obtaining $y(0) = h^{-1}(y)eX, yeX_{i(m)}$.

REFERENCES

[1] Eaves, B. C., "An Odd Theorem", Proceedings of the American Mathematical Society, 26 (1970), 509-513.

[2] Eaves, B. C., "Computing Kakutani Fixed Points", SIAM Journal of Applied Mathematics, 21 (1971), 236-244.

[3] Eaves, B. C., "Homotopies for Computation of Fixed Points", Mathematical Programming, 3 (1972), 1-22.

[4] Eilenberg, S. and Montgomery, D., "Fixed Point Theorems for Multi-Valued Transformations", American Journal of Mathematics, 68 (1946), 214-222.

[5] Hansen, T. and Scarf, H., "On the Applications of a Recent Combinatorial Algorithm", Cowles Foundation Discussion Paper 272 (1969).

[6] Kuhn, H. W., "Simplicial Approximation of Fixed Points", Proceedings of the National Academy of Sciences, 61 (1968), 1238-1242.

[7] Scarf, H., "The Approximation of Fixed Points of a Continuous Mapping", SIAM Journal of Applied Mathematics, 15 (1967), 1328-1343.

Investigations Into the Efficiency
of the Fixed Point Algorithms

R. Saigal

ABSTRACT

In this paper, we study the effect of different labelings and triangulations on the efficiency of the fixed point
algorithms. In particular, we show that the convergence and
accuracy are functions of the particular labeling employed.
In addition two triangulations are studied and theoretically
speculated to be different. Empirical results on implementing these two triangulations in a fixed point algorithm,
confirm this difference.

0. Introduction

Our purpose herein is to study algorithms which compute a fixed point of a continuous mapping f from R^n (the n-dimensional Euclidean space) to R^n ; i.e., a point x in R^n such that $f(x) = x$. Two notable algorithms in this regard are [1], [6].

These algorithms, starting with a constant map $f^0(x) = x^0$, deform f^t as $t \to \infty$ to $f^\infty = f$, and follow the path x^t of fixed points of f^t . If $x^t \to x^\infty$ on a se - quence of t's tending to infinity, then x^∞ is a fixed point of f . In the contrary case, the algorithm fails. Some properties of f which guarantee that these algorithms will not fail are given in [1].

The aim of this paper is to study the effect of the la- beling and the triangulation on the efficiency of the algo- rithm. It contains three sections. §1 studies the effect of labeling on the algorithm, §2 studies the convergence proper- ties of the algorithm, and §3 studies the effect of the triangulation.

1. Labelings

Given a triangulation K of R^n , [8], with the set of vertices $K^0 \subset K$, and an arbitrary set L , a labeling of the triangulation K is a function

$$\ell: \overset{o}{K} \to L .$$

This labeling function ℓ is designed in such a manner that certain simplexes in K become distinguished, and are related to the fixed points of a mapping f .

1.1 Integer Labeling

Given a mapping $f: R^n \to R^n$ by <u>integer</u> <u>labeling</u> we mean that $L = \{1,2,\ldots,n+1\}$. The integer labeling we study here is:

$$\ell(x) = \begin{cases} i & \text{if } f_i(x) > x_i \text{ and } f_j(x) \leq x_j \text{ for all } j < i \\ n+1 & \text{if } f_j(x) \leq x_j \text{ for all } j = 1,\ldots,n . \end{cases}$$

on the vertices x in $\overset{o}{K}$. The distinguished simplexes in this case are called completely labeled, and are simplexes in which each vertex gets a different label. Fixed point algorithms are designed to find such completely labeled sim- plexes. These completely labeled simplexes are related to fixed points in the following manner:

Given a uniformally continuous mapping $f: R^n \to R^n$, and $\varepsilon > 0$, we say δ is determined by ε and the uniform con- tinuity of f whenever $\|x-y\| \leq \delta \Rightarrow \|f(x)-f(y)\| \leq \varepsilon$ (here $\|\cdot\|$ denotes the max norm).

<u>Theorem 1.1</u>: [Jeppson, 2] Let $f: R^n \to R^n$ be uniformally con- tinuous on R^n , and $\varepsilon > 0$ be given. Let K be a triangu- lation of R^n with diameter (mesh) δ , where δ is deter- mined by ε and the uniform continuity of f . Let

$\ell : K^o \rightarrow L$ be the integer labeling defined above. Then for any completely labeled simplex $\sigma = (v^1, \ldots, v^{n+1}) \in K$ we have

$$x \in \sigma \Rightarrow \| f(x) - x \| \leq \varepsilon + \delta \quad .$$

Proof: From the definition of ℓ we have

$$f_j(x) - x_j = f_j(x) - f_j(v^{n+1}) + f_j(v^{n+1}) - v_j^{n+1} + v_j^{n+1} - x_j$$

$$\leq f_j(x) - f_j(v^{n+1}) + v_j^{n+1} - x_j \leq \varepsilon + \delta \quad .$$

Similarly

$$f_j(x) - x_j = f_j(x) - f_j(v^j) + f_j(v^j) - v_j^j + v_j^j - x_j$$

$$\geq f_j(x) - f_j(v^j) + v_j^j - x_j \geq -\varepsilon - \delta \quad .$$

Hence $\| f(x) - x \| \leq \varepsilon + \delta \quad .$

1.2 Vector Labeling

Given a mapping $f : R^n \rightarrow R^n$, and $L = R^n$, by a vector labeling we mean the function

$$\ell(x) = f(x) - x$$

on the vertices x in K^o . The distinguished simplexes in this case are also called completely labeled, and are simplexes $\sigma = (v^1, v^2, \ldots, v^{n+1})$ in K such that $0 \in$ convex hull $\{\ell(v^1), \ell(v^2), \ldots, \ell(v^{n+1})\}$, i.e., the following system of equations

$$\sum_{i=1}^{n+1} \lambda_i \ell(v^i) = 0$$

$$\sum_{i=1}^{n+1} \lambda_i = 1$$

$$\lambda_i \geq 0 \quad , \qquad i = 1,\ldots,n+1$$

has a solution.

    If σ is such a completely labeled simplex one defines

$$x^* = \sum_{i=1}^{n+1} \lambda_i v^i$$

to be an approximate fixed point of f . The justification for calling x^* an approximate fixed point of f is the following:

<u>Theorem 1.2</u>: Let $f: R^n \to R^n$ be uniformly continuous on R^n , and let $\varepsilon > 0$ be given. Let K be a triangulation of R^n with mesh δ where δ is determined by ε and the uniform continuity of f . Let $\ell:K^o \to L$ be the vector labeling defined above. Then for any completely labeled simplex $\sigma = (v^1,\ldots,v^{n+1}) \in K$ and any approximate fixed point

x^* є σ we have

$$\| f(x^*) - x^* \| \leq \varepsilon$$

Proof:

Since x^* є σ $= (v^1, \ldots, v^{n+1})$ we have

$$-\varepsilon \leq f_j(x^*) - f_j(v^i) \leq \varepsilon \qquad \text{for all } i = 1, \ldots, n+1 \quad .$$

Hence

$$-\varepsilon \leq f_j(x^*) - \sum_{i=1}^{n+1} f_j(v^i) \lambda_i \leq \varepsilon \quad .$$

But

$$\sum_{i=1}^{n+1} \lambda_i f_j(v^i) = \sum_{i=1}^{n+1} \lambda_i v_j^i = x_j^* \quad ,$$

and the result follows.

The result of Theorem 1.2 can be considerably improved if f is assumed to be twice continuously differentiable with f_j , j=1,...,n having bounded second derivatives; i.e., there exists an $\alpha > 0$ such that for j = 1,...,n

$$|u^T H_j(x)u| \leq \alpha \|u\| \qquad \text{for all} \quad x \quad \text{and} \quad u \quad \text{in} \quad R^n$$

where $H_j(x)$ is the Hessian of f_j at x .

Theorem 1.3: Let $f: R^n \to R^n$ be twice continuously differential with the f_j , $j=1,\ldots,n$ having bounded second derivatives. Let K be a triangulation of R^n with mesh δ .
Let $\ell:K^o \to L$ be the vector labeling defined above. Then
for any completely labeled simplex $\sigma = (v^1,\ldots,v^{n+1}) \in K$
and any approximate fixed point $x^* \in \sigma$ we have

$$\|f(x^*) - x^*\| \leq \frac{1}{2}\alpha\delta^2$$

Proof:

For each j and $i=1,\ldots,n+1$ we have

$$f_j(v^i) = f_j(x^*) + (v^i - x^*)^T \nabla f_j(x^*) + \frac{1}{2}(v^i - x^*)^T H_j(u^i)(v^i - x^*)$$

where u^i is some point on the line segment joining v^i and
x^* . Substituting into the definition of x^* we get

$$x_j^* = \sum_{i=1}^{n+1} \lambda_i f_j(v^i) = f_j(x^*) + \left(\sum_{i=1}^{n+1} \lambda_i v^i - x^*\right)^T \nabla f_j(x^*)$$

$$+ \frac{1}{2}\sum_{i=1}^{n+1} \lambda_i (v^i - x^*)^T H_j(u^i)(v^i - x^*)$$

Thus

$$\left| x_j^* - f_j(x^*) \right| \le \frac{1}{2} \sum_{i=1}^{n+1} \lambda_i \left| (v^i - x^*)^T H_j(u^i)(v^i - x^*) \right|$$

$$\le \frac{1}{2} \alpha \delta^2$$

and the result follows.

2.   Convergence Properties

The fixed point algorithms produce a sequence of approximate fixed points x^k, $k = 1, 2, \ldots$ on a sequence of grids of diameter δ^k, $k = 1, 2, \ldots$ such that the sequence δ^k converges to zero. It can be readily shown (see §1) that all cluster points of the sequence x^k are fixed points of f.

In this section we will show that the rate of convergence of the sequence x^k is, at least, linear. For this we make the following assumptions

2.1   The sequence x^k converges to x^∞

2.2 f is twice continuously differentiable.

2.3   The matrix $D(f-I)(x^\infty)$ is nonsingular. (Here Dh is the Jacobian matrix of the mapping h and I is the identity map).

2.4   The Hessian matrix H_j of f_j , $j = 1,\ldots,n$,
has the property that $|u^T H_j(x)\, u| \leq \alpha\, \|u\|^2$ for
all u and x .

We are now ready to prove

<u>Theorem 2.1</u>:   Let the assumptions 2.1-2.4 hold on a se-
quence of approximate fixed points generated by a fixed
point algorithm, and let $\|x^k - f(x^k)\| < \theta_k$.   Then, for
large enough k , there is a $\rho > 0$ such that

$$\|x^k - x^\infty\| \leq \rho \theta_k \quad .$$

<u>Proof</u>:
Taking the second order Taylor's expansion of f_j
about x^∞ , we get

$$f_j(x^k) = f_j(x^\infty) + \nabla f_j(x^\infty)\,(x^k - x^\infty) + \frac{1}{2}(x^k - x^\infty)^T H_j(u)(x^k - x^\infty) \; .$$

Since x^∞ is a fixed point of f , rewriting we get

$$-\frac{1}{2}(x^k - x^\infty)^T H_j(u)(x^k - x^\infty) - x_j^k + f_j(x^k) = -x_j^k + x_j^\infty$$

$$+ (x^k - x^\infty)\nabla f_j(x^\infty)$$

or

$$\frac{1}{2}\alpha\| x^k - x^\infty \|^2 + \theta_k \geq | (\nabla f_j(x^\infty) - e_j)(x^k - x^\infty) |$$

or

$$\frac{1}{2}\alpha\| x^k - x^\infty \|^2 + \theta_k \geq \| D(f-I)(x^\infty)(x^k - x^\infty) \| \quad .$$

Since $A = D(f-I)(x^\infty)$ is nonsingular, there is a $\theta > 0$ such that

$$\| A\ u \| \geq \theta \| u \| \quad .$$

Hence, we obtain

$$\theta_k \geq \theta \| x^k - x^\infty \| - \frac{1}{2}\alpha \| x^k - x^\infty \|^2 \quad .$$

Since the sequence x^k converges to x^∞, for large enough k the quadratic term will be neglibible, giving the result with $\rho = \frac{2}{\theta}$.

As is seen in §1, θ_k can be computed as a function of δ_k and ε_k . Its exact form depends on the labeling and assumptions about the function. In case we were using vector labels, using Theorem 1.3, we obtain the

Corollary 2.1. Under the assumption 2.1-2.4, and that vector labeling is employed in the algorithm, there is a

$\rho > 0$ such that for sufficiently large   k   ,

$$\| x^k - x^\infty \| \leq \frac{1}{2} \rho \alpha \delta_k^2 \quad .$$

Proof:

Follows from the fact that Theorem 1.3 implies

$\theta_k = \frac{1}{2} \alpha \delta_k^2$, and thus the result follows from Theorem 2.1.

3.   Triangulations

In this section we study the effect of triangulations on the efficiency of the fixed point algorithm. The results we present here are quite incomplete, and further results will be reported elsewhere.

We restrict our attention to the algorithms of Merrill [6] and Eaves and Saigal [1]. Both these algorithms employed the following triangulation of R^n , we call   H   , which is generated by picking an arbitrary positive real number   D   . The vertices of this triangulation, H^0 , are the set of all points in R^n whose coordinates are integer multiples of   D   . Each simplex in   K   is uniquely represented by a pair (v, π) where $v \in H^0$ and π is a permutation of $\{1,2,\ldots,n\}$. Given the pair (v, π) , the vertices of the simplex are generated as follows:

Let

$$Q = \begin{bmatrix} -D & 0 & 0 & \cdots & 0 & 0 \\ D & -D & 0 & \cdots & 0 & 0 \\ 0 & D & -D & \cdots & 0 & 0 \\ \cdot & \cdot & \cdot & \cdot & \cdot & \cdot \\ \cdot & \cdot & \cdot & \cdot & \cdot & \cdot \\ \cdot & \cdot & \cdot & \cdot & \cdot & \cdot \\ 0 & 0 & 0 & \cdots & -D & 0 \\ 0 & 0 & 0 & \cdots & D & -D \end{bmatrix}$$

Then $(v, \pi) = (v^1, v^2, \ldots, v^{n+1})$ where

$$v^1 = v$$

and

$$v^{i+1} = v^i + Q_{\pi_i} , \qquad i = 1, \ldots, n .$$

where Q_{π_i} is the π_i^{th} column of Q .

This triangulation, when restricted to the unit cube $C = \{x \in R^n \mid 0 \le x_i \le 1, \qquad i = 1, \ldots, n\}$, produces $n!$ simplexes in C . For the case $n = 3$, these simplexes are shown in the Figure 3.1.

Given that the vertices of C are labeled as in the Figure 3.1, the six simplexes in this case are the convex hull of the points in the sets $\{1,2,4,5\}$, $\{3,6,7,8\}$,

$\{2,3,4,8\}$, $\{2,3,6,8\}$, $\{2,4,5,8\}$, $\{2,5,6,8\}$.

Figure 3.1

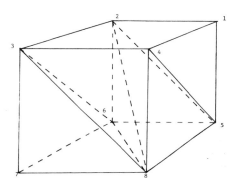

There is another triangulation of R^n , Kuhn [4], which can be used instead of H . We call this triangulation I , and it is generated as follows. Given a positive real number D , the vertices I^0 of this triangulation are all vectors in R^n whose coordinates are integer multiples of D (as in H). Each simplex in I has a unique representation (v,π) where $v \in I^0$ and π is a permutation of $\{1,\ldots,n\}$. Given the pair (v,π) the vertices of this triangulation are generated as follows:

Let

$$
P = \begin{bmatrix}
-D & 0 & 0 & \cdots & 0 \\
0 & -D & 0 & \cdots & 0 \\
0 & 0 & -D & \cdots & 0 \\
\cdot & \cdot & \cdot & \cdot & \cdot \\
\cdot & \cdot & \cdot & \cdot & \cdot \\
\cdot & \cdot & \cdot & \cdot & \cdot \\
0 & 0 & 0 & \cdots & -D
\end{bmatrix}
$$

Then $(v, \pi) = (v^1, v^2, \ldots, v^{n+1})$ where

$$
v^1 = v
$$

and

$$
v^{i+1} = v^i + P_{\pi_i} \qquad i = 1, \ldots, n \quad .
$$

where P_{π_i} is the π_i column of P .

This triangulation, when restricted to the unit cube C also results in $n!$ simplexes. For the case $n = 3$, these simplexes are shown in Figure 3.2 (contrast with Figure 3.1)

Figure 3.2

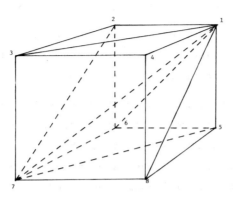

Given that the vertices of C are labeled in Figure 3.2, the six simplexes of this triangulation of C are {1,3,4,7}, {1,2,3,7} , {1,4,8,7}   {1,5,8,7} , {1,6,5,7} , {1,2,6,7} .

Another triangulation of R^n which could have significant impact on the efficiency of the fixed point algorithm is the one studied by Mara [5], who produced a four- and five-dimensional analogues of it. The generalization to arbitrary n is not known. In the case n = 3 , it produces five simplexes shown in Figure 3.3

Figure 3.3

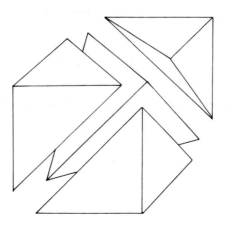

We now study the effect of the triangulations H and
I on the efficiency of the fixed point algorithms. Note
that the amount of work done to go from x^k to x^{k+1} (in
the sequence studied in §2) is controlled by the triangulation. As the fixed point algorithm pivots through these simplexes in R^n it passes through many cubical regions of the
space. Thus, an estimate of the effort expended in proceeding through the space may be measurable by studying the effort required to pass through a cubical region of the space.
This then allows us to localize our study to the effect of
different triangulations of C on the effort required to
pivot through C . In both H and I , the worst case is
to go through n! of these simplexes. We now define a measure which seems to explain the behavior of the algorithm
better than n! . To do this, we restrict our attention to
the cube C and its induced triangulation by H and I ,
which we also refer to as H and I respectively.

    We now make explicit our notation:

3.1 <u>Boundary Facets of a Triangulation of C</u>: are the facets
of the triangulation of C which lie on the boundary of C .

3.2 <u>Simple Path in the Triangulation</u>: is a sequence

$$\tau^0, \sigma_1, \tau_1, \ldots, \sigma_k, \tau_k$$

of distinct facets τ_i, i = 0,...,k and distinct simplexes
σ_i, such that τ_{i-1} and τ_i are facets of σ_i, i = 1,...,k.
We say that this is a path between τ_0 and τ_k and has
length k .

3.3 Minimal Path between Boundary Facets: Given a pair of
boundary facets τ and $\bar{\tau}$, by $\ell(\tau,\bar{\tau})$ we represent the
length of the minimal length simple path between τ and $\bar{\tau}$
in the triangulation.

3.4 Diameter (Dia) of the Triangulation: is the maximal of
$\ell(\tau,\bar{\tau})$ between any pair of boundary facets τ and $\bar{\tau}$,
i.e.,

$$Dia = \max_{\tau,\bar{\tau}} \quad (\tau,\bar{\tau})$$

We hypothesize that the diameter of the induced triangulation of C controls the efficiency of the algorithm.
To test this, both the triangulations H and I were implemented in Merrill's algorithm. Various problems were solved
using both of them, starting from the same point in each
algorithm. The diameters for both these triangulations are
relatively simple to compute. They are:

Triangulation	H	I
Diameter	$\geq 0(n^3)$	$\dfrac{n(n-1)}{2}$

The results obtained when solving four nonlinear programming problems by these algorithms are summarized in
Table 3.1. The dimension of the problems were 5, 6, 8 and
15 respectively.

Table 3.1

Dimension n	# Starting Points	Triangulation	Total Pivots for all Starting Points	ratio = $\frac{\text{Pivots H}}{\text{Pivots I}}$	ratio for each run
5	8	I	11,166	1.56	1.12-2.23
		H	17,416		
6	10	I	14,964	2.94	1.66-4.65
		H	41,095		
8	6	I	8,463	3.26	2.15-4.30
		H	27,590		
15	1	I	990	3.59	
		H	3,570		

Similar results have been found by Wilmuth [9] when triangulations H and I are compared in the Eaves and Saigal [1] algorithm. He solved the three economic equilibrium problems of Scarf [7], of dimensions 4, 7 and 9 respectively.

These triangulations were also tested on a problem supplied by Kellogg [3]. This problem has a 20 and a 80 dimensional version. Triangulations H and I were, as before, implemented in Merrill's algorithm. The results for the 20 dimensional problem are summarized in Table 3.2

Table 3.2

20 Dimensional Fixed Point Problem

Run	Starting Point	Pivots H	Pivots I	$\dfrac{\text{Pivots H}}{\text{Pivots I}}$
1	000.....0 0	586	138	4.25
2	111.....1 1	1446	234	6.18
3	-1 1 -1.....-1 1	562	318	1.77
4	-1-1-1.....-1-1	2292	112	20.46
5	-1-1-1-1-1 11111 -1-1...1	706	338	2.09
6	1 11 -1-1-1 111...1	628	580	1.08
7	11 -1 11 -1 ... -1	262	310	0.85
8	1 -1 1 -1 ... -1	588	292	2.01

The results from the 80 dimensional problem are quite dramatic. Starting with the point $(1,1,\ldots,1,1)$ triangulation I solved the problem in 2341 pivots (108.95 secs of CPU time in IBM 370/68) while even after 49,000 pivots (40 mins of CPU time) triangulation H had not converged to the

first level. The one run that both solved was with the
starting point (1,1,-1,1,1,-1,...,1). Triangulation H
took 14,988 pivots, while triangulation I took 4349, a ratio of 3.42. The other runs with triangulation I are tab-
ulated in Table 3.3

Table 3.3

Run	Starting Point	# Pivots with I
1	0,0,0,...	1413
2	1,1,1,...	2341
3	1,-1,1,...	4345
4	-1,-1,1 1,-1,-1,...	4595
5	0,-1,0,-1,...	2649
6	1 1 -1,1 1 -1,...	4349
7	-1,-1,-1,...	1393
8	-1,1,-1,1,...	4571

REFERENCES

[1] Eaves, B. C., and Saigal, R., "Homotopies for Computa-
     tion of Fixed Points on Unbounded Regions," Math. Prog.,
     3, 1972.

[2] Jeppson, M., "A Search for the Fixed Points of a Contin-
     uous Mapping," Mathematical Topics in Economic Theory
     and Computations, SIAM, Philadelphia, 1972.

[3] Kellogg, R. B., private communication.

[4] Kuhn, H. W., "Some Combinatorial Lemmas in Topology,"
     IBM Journal of Research and Development, 4, 1960.

[5] Mara, P. S., "Triangulations of a Cube," M.S. Thesis,
     Colorado State University, Fort Collins, Colorado, 1972.

[6] Merrill, O. H., "Applications and Extensions of an Algo-
     rithm that Computes Fixed Points of Certain Upper Semi-
     continuous Point to Set Mappings," Ph.D. Thesis, Univer-
     sity of Michigan, Ann Arbor, 1972.

[7] Scarf, H. Computation of Economic Equilibrium, Yale
     University Press, New Haven, 1973.

[8] Spanier, E. H. Algebraic Topology, McGraw-Hill, New
     York, 1966.

[9] Wilmuth, R. J., private communication.

Almost-Complementary Paths in the
Generalized Complementarity Problem

John Freidenfelds

ABSTRACT

If K is a closed convex cone in R^n , K^+ is its
polar cone and w is a continuous mapping from K into R^n,
then the generalized complementarity problem is to find x
such that $x \in K, w(x) \in K^+$, and the inner product $xw(x)$
$= 0$. Using a fixed-point algorithm it is shown that there
is a connected set of almost-complementary points which joins
the origin to either a complementary point or to infinity.
This is a generalization of the almost-complementary path
generated by Lemke for the linear complementarity problem
$(K = R^n_+$ and w affine).

1.   Introduction

The complementarity problem as originally posed
[17,18,4]   , henceforth called the linear complementarity
problem, is to find $x \in R^n$ such that

$$x \geq 0 \quad ,$$

$$Mx + q \geq 0 \quad , \tag{1.1}$$

and $x(Mx+q) = 0 \quad ,$

where $M \in R^{n \times n}$ and $q \in R^n$. Such a problem arises in a
variety of contexts. In other contexts the affine function
Mx+q is replaced by some nonlinear function $w(x)$ [3,13] .
A further generalization [10] is to find $x \in R^n$ such that

$$x \in K \quad ,$$

$$w(x) \in K^+ \quad , \tag{1.2}$$

and $xw(x) = 0 \quad ,$

where K is a convex cone in R^n and K^+ it polar cone.

The linear complementarity problem has been attacked
largely by elementary algebraic means [17,18,4] while the

nonlinear version has often been pursued with fixed point
theorems [13,10,5]. Using Kakutani's [12] fixed point theorem for upper-semicontinuous point-to-set maps, Karamardian
[14,15] has established the existence of solutions of (1.2)
for certain classes of functions. Eaves [5] has established
the existence, in the nonlinear case, of almost-complementary
sets of points analogous to the almost-complementary paths
[17,18] used in solving (1.2). His proof depended on a theorem of Browder [2] (a parametric version of Brouwer's fixed
point theorem).

   Meanwhile, Scarf and others [6,7,8,11,16,19,20,21] stimulated by the algebraic methods of solving the linear complementarity problem, have developed elementary "algebraic"
algorithms for approximating fixed points of Brouwer and
Kakutani maps. It is demonstrated here that their fixed-
point algorithms are more powerful than the Brouwer or
Kakutani fixed point theorems. The fixed-point algorithms
can, in fact, be used to prove Browder's theorem in R^n and
hence to establish the existence of almost-complementary
"paths" in the general nonlinear case of (1.2). The nature
of such "paths" (actually, connected sets of almost-complementary points) can be used, for example, to establish some
existence theorems of Karamardian [14,15] for the nonlinear
complementarity problems.

2.   The Fixed-Point Algorithm

   Most of the original applications of the fixed point
algorithm boil down [6] to finding fixed points whose existence is guaranteed by the following.

(2.1) <u>Theorem</u> (Kakutani [12]) Let C be a compact convex

subset of R^n , C* the collection of convex subsets of C ,

and f:C → C* an upper semi-continuous map. Then f has a

fixed point.

Of interest here are the more recent, algorithmic or

"constructive" proofs [6,7,8,11,16,19,20,21]. We shall

sketch a version which is very close to one used by Kuhn [16]

and Eaves [7].

Assume C has an interior point, c , relative to R^n

(otherwise, work in a lower dimensional space). Approximate

the points of R^n with the points

$$\pi \equiv \{x \in R^n \mid x_i = N_i/N \qquad \text{where}$$

$$N_i \text{ is an integer,} \qquad i = 1,\ldots,n\} \quad , \quad (2.2)$$

where the integer N > 0 determines the grid size or refine-

ment of the approximation. To generate a simplicial subdivi-

sion of R^n using the points of π as vertexes, define the

n x n matrix.

$$U \equiv \frac{1}{N} \begin{bmatrix} -1 & 0 & & 0 & 0 \\ +1 & -1 & & \cdot & \cdot \\ 0 & +1 & & \cdot & \cdot \\ \cdot & 0 & & 0 & \cdot \\ \cdot & \cdot & & -1 & 0 \\ 0 & 0 & & +1 & -1 \end{bmatrix} \qquad (2.3)$$

and let $U(i)$ designate its ith column. Let $\{v^o,\ldots,v^n\}$ ε π be vertices of a full-dimensional simplex if and only if there is some permutation $\gamma = (\gamma_1,\ldots,\gamma_n)$ of the integers $1,\ldots,n$ such that

$$v^i = v^{i-1} + U(\gamma_i) \qquad i = 1,\ldots,n \quad . \qquad (2.4)$$

Thus every full-dimensional simplex is given by some pair (v^o,γ) . When we wish to show explicit dependence on N we designate it by $(v^o,\gamma)^N$. Figure 1 illustrates for $n = 2$, $N = 3$. Determination of adjacent full-dimensional simplexes becomes a trivial numerical procudure with such a set-up. Given a simplex (v^o,γ) , to determine the adjacent simplex which shares the fact opposite vertex v^i , generate the new simplex $(\bar{v}^o,\bar{\gamma})$ as indicated in the following table.

<u>Table 1</u>

	$\bar{v}^0 =$	$\bar{\gamma} =$
$i = 0$	$v^o + U(\gamma^1)$	$(\gamma^2,\ldots,\gamma^n,\gamma^1)$
$1 \leq i \leq n-1$	v^o	$(\gamma^1,\ldots,\gamma^{i+1},\gamma^i,\ldots,\gamma^n)$
$i = n$	$v^o - U(\gamma^n)$	$(\gamma^n,\gamma^1,\ldots,\gamma^{n-1})$

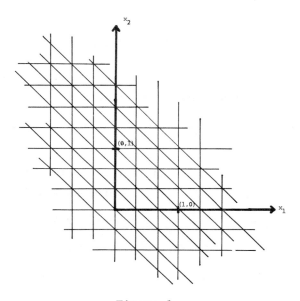

Figure 1

We now describe an algorithm which finds approximate
fixed points for Theorem 2.1. They are approximate fixed
points in the sense that if one is found for each N , then
as $N \to \infty$, the accumulation points of this sequence of approximate fixed points are fixed points of f . To define
an approximate fixed point, we introduce the following labeling of π . For $x \in \pi$ let

$$
g(x) \equiv
\begin{cases}
y - x & \text{where} \quad y \in f(x), \quad \text{if} \quad x \in C \\
\\
c - x & \text{otherwise.}
\end{cases}
\qquad (2.5)
$$

Given any $d \in R^n$, $d \neq 0$ (called a <u>direction</u> <u>vector</u>), we
say that a simplex (v^o, γ) is <u>almost-completely</u> <u>labeled</u> (ab-
breviated acl) with respect to d if and only if the fol-
lowing system has a solution

$$\sum_{i=0}^{n} \lambda_i g(v^i) + \theta d = 0$$

$$\sum_{i=0}^{n} \lambda_i + \theta = 1 \qquad\qquad (2.6)$$

$$\lambda_i \geq 0 \qquad i = 0, \ldots, n, \qquad \theta \geq 0 ;$$

and is <u>completely</u> <u>labeled</u> if (2.6) has a solution with
$\theta = 0$. It follows easily from the convexity of C and
the upper semicontinuity of f that all the points of completely labeled simplexes are approximate fixed points
(Eaves [6] has shown, in fact, that the point $\sum_{i=0}^{n} \lambda_i v^i$,
where λ_i satisfy (2.6) with $\theta = 0$ is the exact fixed-
point of a certain affine approximation of f).

To prove Theorem 2.1 we now demonstrate that a completely labeled simplex can always be found by generating a sequence of adjacent, almost-completely labeled (acl) simplexes which must inevitably lead to one which is completely
labeled. Start with an arbitrary simplex, $\sigma^o = (v^o, \gamma)$,
which is entirely outside of C , and let $\bar{c}$ be interior
to the simplex σ^o . Define direction vector $d \equiv \bar{c} - c$.

Then σ^o is acl (let $\theta = \frac{1}{2}$, $\lambda_i = \mu_i/2$, where

$$\bar{c} = \sum_{i=0}^{n} \mu_i v^i \quad , \quad \mu_i \geq 0 \quad , \quad \sum_{i=0}^{n} \mu_i = 1 \quad).$$ Furthermore, any

simplex outside of C is acl if and only if it contains points on the ray originating at c and going through $\bar{c}$.

To see this, note that if $v^i \not\in C$ $i = 0,\ldots,n$, then $g(v^i) = c - v^i$ and (2.6) can be written

$$c + \delta d = \sum_{i=0}^{n} \alpha_i v^i$$

$$\sum_{i=0}^{n} \alpha_i = 1 \qquad\qquad (2.7)$$

$$\delta \geq 0, \quad \alpha_i \geq 0, \qquad i = 0,\ldots,n .$$

In fact, the ray $r \equiv \{x \in R^n | x = c + \delta d, \delta \geq 0\}$ is entirely covered outside of C by a unique* sequence of adjacent, acl simplexes. This sequence can be generated at will using Table 1 and Equation 2.6 as follows. Suppose (v^o, γ) is an acl simplex outside of C . Then the linear system (2.6) has a solution, and $\theta > 0$ in any solution. If it has a solution, it must have a basic solution (i.e., one using at most $n+1$ variables since there are $n+1$ equations). Let

---

*A slight perturbation of c or $\bar{c}$ may be necessary to remove degeneracies.

$\lambda_p = 0$ be the only nonbasic variable (there are $n+2$ variables altogether) and assume c or $\bar{c}$ has been perturbed if necessary so that all other $\lambda_i > 0$. Introduce λ_p into the basis. Some other, uniquely determined variable, say λ_q , will become nonbasic. Using Table 1, we can determine the simplex adjacent to σ^o which shares the face opposite v^p ; and also the one which shares the face opposite v^q . These are two uniquely determined acl simplexes adjacent to σ^o . If we move to one of these adjacent simplexes, we can apply the same argument to say that it, in turn, has two adjacent acl simplexes -- the one we just came from and another. Thus, we can continue to move from acl simplex to uniquely determined adjacent acl simplex. Starting from σ^o , there are thus two unique sequences of adjacent, acl simplexes. One of these leads away from C along the ray r , and the other, more interesting sequence, leads toward C . The situation is illustrated in Figure 2 on page 10 for $n = 2$. We now consider what happens if we continue generating the sequence into C by alternately performing a linear programming pivot in (2.6) to determine the next nonbasic variable, and using Table 1 to generate the next adjacent acl simplex. We have shown that as long as $\theta > 0$ in the solution of (2.6), the process can go on. Also, since each acl simplex generated has exactly two adjacent acl simplexes none can be repeated unless the whole sequence is repeated, but the sequence cannot be repeated since all acl simplexes outside of C lie on the ray r . Thus, the sequence must terminate finitely with a solution of (2.6) in which $\theta = 0$ -- i.e., a completely labeled simplex.

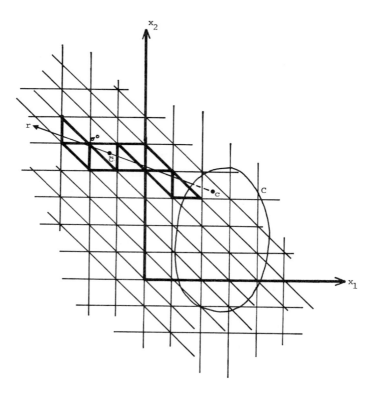

Figure 2

3. Almost-Complete Points

We have shown that the fixed-point algorithm can be used to compute approximate fixed points. In this section we examine the entire sequence of acl simplexes generated by that algorithm.

With f, C, c and d as defined in Section 2, define the set of almost-complete points as

$$
A_d \equiv \left\{ x \in R^n \;\middle|\;
\begin{array}{lll}
0 \in \text{conv } [f(x)-x,d] & \text{if} & x \in \text{interior } C, \\
0 \in \text{conv } [f(x)-x,d,c-x] & \text{if} & x \in \text{boundary } C, \\
0 \in \text{conv } [d,c-x] & \text{if} & x \notin C
\end{array}
\right\}
$$

(3.1)

and the set of complete points, $A \subset A_d$, as

$$
A \equiv \left\{ x \in R^n \;\middle|\;
\begin{array}{lll}
0 \in \text{conv } [f(x)-x] & \text{if} & x \in \text{interior } C \\
0 \in \text{conv } [f(x)-x,c-x] & \text{if} & x \in \text{boundary } C
\end{array}
\right\}
$$

(3.2)

where conv [·] is the convex hull and $f(x) - x \equiv \{z \in R^n \;|\; z = y - x, y \in f(x)\}$. Several consequences follow quickly from the definitions.

(i) A is just the set of fixed points of f ,
     $\{x \in C \mid x \in f(x)\}$;

(ii) $A \subset A_d$;

(iii)   points in the acl simplexes converge to A_d as
        $N \to \infty$;

(iv)   points in the completely labeled simplexes con-
        verge to A as $N \to \infty$;

(v)   outside of the set C, A_d coincides with the
        ray $r = \{x \in R^n \mid x = c + \delta d, \; \delta > 0\}$;

(vi)   both A_d and A are closed sets in R^n.

We indicate a proof only for (iii). Let $(v^o, \gamma)^N$ be a sequence of acl simplexes as $N \to \infty$. Let γ_i^N, θ^N be the corresponding solutions of (2.6). Choose a subsequence such that $(v^i)^N \to \bar{x}$, $\lambda_i^N \to \bar{\lambda}_i$, $\theta^N \to \bar{\theta}$. It is easy to check that equations (2.6), in the limit, establish $0 \in \text{conv } [f(\bar{x}) - \bar{x}, d]$ if $\bar{x} \in$ interior C, etc., as required for $\bar{x}$ to be an element of A_d.

We are now in a position to prove an extension of Theorem (2.1).

(3.3)   <u>Theorem</u>   The connected component of A_d which contains the ray r outside of C also contains a fixed point (a point of A).

<u>Proof</u>: We shall use a concept called ε-chaining (our definitions differ only slightly from those of [1]). Denote for $Q \subset R^n$ and $\varepsilon > 0$, the <u>ball of radius</u> ε about Q as $\beta(Q, \varepsilon) \equiv \{x \in R^n \mid d(x, Q) \leq \varepsilon\}$, where $d(x, Q)$ is the

distance from x to the set Q . Two points a,b $\in$ R^n are
called ε-chained in Q if and only if there exist
$x^1, x^2, \ldots, x^K$ in $\beta(Q,\varepsilon)$ with $d(a,x^1) \leq \varepsilon$, $d(x^i, x^{i+1}) \leq \varepsilon$,
i = 1,...,K-1 , and $d(x^K, b) \leq \varepsilon$. If a and b are
ε-chained in Q for all ε > 0 , then they are said to be
well-chained in Q .

(3.4) <u>Lemma</u> Suppose Q is a compact set in R^n . Then if
a and b are well-chained in Q , they belong to the same
connected component of Q .

<u>Proof</u>: Suppose, contrary to the lemma, the open disjoint
sets U and V with a $\in$ U and b $\in$ V cover Q . Then
the closed disjoint sets U' $\equiv$ {x $\in$ Q | x $\notin$ V} and
V' $\equiv$ {x $\in$ Q | x $\notin$ U} also cover Q and a $\in$ U' , b $\in$ V' .
Let $\varepsilon_o = \frac{1}{3} d(U',V') > 0$. Clearly a and b cannot be
ε_o-chained in Q , and the lemma is proved.

To prove the theorem, observe that the fixed-point algorithm generates a sequence of points of $\pi \subset R^n$ (each new
adjacent acl simplex produces another point), $(P_o^N, P_1^N, \ldots, P_\ell^N)$,
where $(P_o^N, \ldots, P_n^N)$ are the vertexes of the initial acl simplex which contains $\bar{c}$; $(P_i^N, \ldots, P_{i+n}^N)$ for i = 1,...,ℓ-n-1
are the vertexes of intermediate acl simplexes; and
$(P_{\ell-n}^N, \ldots, P_\ell^N)$ are the vertexes of the final completely
labeled simplex. Of course the distance between successive
points P_j^N, P_{j+1}^N is at most $\sqrt{2}/N$. Since acl simplexes

converge to A , we can, for any $\varepsilon > 0$, choose M sufficiently large that for all $N \geq M$, $P_o^N \in \beta(\bar{c},\varepsilon)$, $P_j^N \in \beta(A_d,\varepsilon)$, $j = 1,\ldots,\ell-1$, $P_\ell^N \in \beta(A,\varepsilon)$; and such that the distance between successive points, $P_o^N, P_1^N, \ldots, P_\ell^N$, is less than ε . Thus $\bar{c}$ is well-chained in A_d to some point of A and the theorem is proved.

4.   Browder's Theorem in R^n

In this section we use Theorem (3.3) to give an elementary proof of a "parameterized" version of the Kakutani Theorem (2.1).

(4.1)   Theorem (Browder [2])   Let C and C^* be as in Theorem (2.1) and let $F:C \times [0,1] \to C^*$ be an upper semicontinuous map. Then there exists a closed connected set $T \subset C \times [0,1]$ which intersects both $C \times \{0\}$ and $C \times \{1\}$ and $x \in F(x,t)$ for all $(x,t) \in T$.

Proof: Define the Kakutani map $\tilde{F}$ in R^{n+1} as $\tilde{F}(x,t) \equiv (F(x,t),0)$. Apply Theorem (3.3) to $\tilde{F}$ with $\tilde{c} = (c,1/2)$ where $c \in$ interior C , and $d = (0,1) \in R^{n+1}$. The situation is illustrated for $n = 2$ in Figure 3. It is straightforward to show that

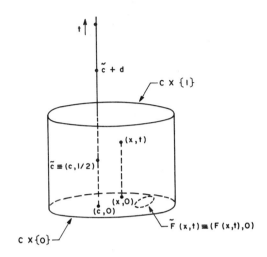

Figure 3

(i)   the ray r enters C x [0,1] through C x {1};

(ii) almost-complete points with $0 \leq t < 1$ are fixed
      points of F ; and

(iii) the only complete points are fixed points in
      C x {0} .

Now, with a little more effort for $t \to 1$, the Theorem follows from Theorem (3.3) (see [9] for details).

5.   The Generalized Complementarity Problem

Let K be a closed, convex cone in R^n , and

$K^+ \equiv \{x \in R^n \mid xy \geq 0 \text{ for all } y \in K\}$, its polar cone.

Assume K is such that there exists some vector, $p \neq 0$

(assume $\|p\| = 1$), and $p \in$ interior K^+ .   Let $w : K \to R^n$

be a continuous function. Denote as the <u>generalized comple-</u>
<u>mentarity problem</u> (GCP) that of finding x , called a
<u>complementary point</u>, such that

$$x \in K, \quad w(x) \in K^+, \quad \text{and} \quad xw(x) = 0 \quad . \quad (5.1)$$

We call x <u>almost-complementary</u> (with respect to p) if
and only if for some $\tau \geq 0$,

$$x \in K, \quad (x(x)+\tau p) \in K^+, \quad \text{and} \quad x(w(x)+\tau p) = 0 \quad . \quad (5.2)$$

we note that the Lemke-Howson formulation [18] is a special
case of this one for w an affine function and $K = R_+^n$.

We shall prove an "almost-complementary path" theorem
for this problem which is a generalization of one due to
Eaves [5] for $K = R_+^n$.

(5.3) <u>Theorem</u> There is a closed, connected set of almost-
complementary points which contains the origin and either
contains a complementary point or is unbounded.

<u>Proof</u>: Let $\bar{A}_p \equiv \{x \in K \mid$ for some τ (not necessarily
≥ 0), $(w(x)+\tau p) \in K^+$ and $x(w(x)+\tau p) = 0\}$. We shall show
that the points of $\bar{A}_p$ are the fixed points of a certain
parameterized Kakutani map so that Theorem (5.3) follows
from Theorem (4.1). Let $t \geq 0$ and $C_t \equiv \{z \in K \mid zp = t\}$.
Noting that for $\alpha > 0$, C_α is a compact, convex set in a

subspace of dimension n-1 , we define the upper semi-continuous mapping $F:C_\alpha \times [0,\alpha] \to C_\alpha^*$ for $t > 0$ as

$$F(z,t) \equiv \frac{\alpha}{t} \frac{argmin}{y \, \epsilon \, C_t} \, yw \left(\frac{t}{\alpha} z \right) , \qquad (5.4)$$

and for $t = 0$ as all of the limit points of (5.4), where

$$\frac{argmin}{y \, \epsilon \, C_t} \, yq \equiv \{ y \, \epsilon \, C_t \mid yq \leq hq \qquad all \qquad h \, \epsilon \, C_t \}. \quad (5.5)$$

It is easy to check that if $z \, \epsilon \, F(z,t)$ then $x \equiv (t/\alpha)z \, \epsilon \, \frac{argmin}{y \, \epsilon \, C_t} \, yw(x)$, so that $xw(x) \leq hw(x)$, all $h \, \epsilon \, C_t$. Thus, letting $\tau \equiv -xw(x)/xp = -xw(x)/t$, we have for any $h \, \epsilon \, C_t$, $h(w(x)+\tau p) = hw(x) + \tau t = hw(x) - xw(x) \geq$ 0 , so $(w(x)+\tau p) \, \epsilon \, K^+$. Also $x(w(x)+\tau p) = 0$, so $x \, \epsilon \, \bar{A}_p$. Since this holds for any $\alpha > 0$, Theorem (4.1) establishes the following.

(5.5) <u>Lemma</u> The connected component of $\bar{A}_p$ which touches the origin, $\bar{T}$, is unbounded.

For any nonzero $x \, \epsilon \, \bar{T}$, let $\tau(x) \equiv -xw(x)/xp$. If $\tau(x) > 0$ for all nonzero $x \, \epsilon \, \bar{T}$, then $\bar{T}$ is an unbounded closed connected set of almost-complementary points which contains the origin. If for some $\bar{x} \, \epsilon \, \bar{T}$, $\tau(\bar{x}) \leq 0$, let $0,x^1,x^2,\ldots,x^n$ be an ε-chain in $\bar{T}$ with $\tau(x^i) > 0$, $i < n$, and $\tau(x^n) \leq 0$. The last point, x^n , is either

$\bar{x}$ or some other x encountered along the way for which $\tau(x) \leq 0$. As $\varepsilon \to 0$, if the sequence of x^n diverges, we again have an unbounded closed connected set of almost-complementary points which contains the origin. Otherwise, since $w(\cdot)$ is continuous, $\tau(\cdot)$ is continuous; and so as $\varepsilon \to 0$, $\tau(x^n) \to 0$. Thus the origin is well-chained in the set of almost-complementary points to a complementary point and the proof of (5.3) is complete.

6. Existence Conditions for the Generalized Complementarity Problem (GCP)

Theorem (5.3) can be used to obtain some sufficient conditions for the GCP to have a solution.

In the first corollary, we shall show that the set of almost-complementary points is bounded, and so by (5.3), there must be a complementary point. We say that a map $M:K \to R^n$ is positively homogeneous of degree r over K if and only if $M(\lambda x) = \lambda^r M(x)$, all $\lambda \geq 0, x \in K$.

(6.1) **Corollary** (Karamardian [15] Let K be a cone in R^n, K^+ its polar, and q_o a given vector in interior K^+ . Let $M:K \to R^n$ be continuous and positively homogeneous of degree r . If the GCP with $w(x) \equiv M(x) + q$ has a unique solution (namely x = 0) for q = 0, and for $q = q_o$, then it has a solution for all $q \in R^n$.

Proof: Let $C_\lambda \equiv \{x \in K \mid xq_o = \lambda\}$, $\lambda \geq 0$. Suppose for every $\lambda > 0$, there is an almost-complementary point

$x_\lambda \in C_\lambda$.   Then

$$\tau_\lambda \equiv \frac{-x_\lambda (M(x_\lambda)+q)}{x_\lambda q_o} \geq 0 \qquad (6.2)$$

and

$$(M(x_\lambda)+q) + \tau_\lambda q_o \in K^+ \quad . \qquad (6.3)$$

Writing $x_\lambda = \lambda x_1^\lambda$, where $x_1^\lambda \in C_1$ and using the positive homogeneity of M , these two conditions can be rearranged into

$$x_1^\lambda M\left(x_1^\lambda\right)+ (\lambda^{-r})x_1^\lambda q \leq 0 \qquad (6.4)$$

and

$$M\left(x_1^\lambda\right) - \frac{x_1^\lambda M\left(x_1^\lambda\right)}{x_1^\lambda q_o}\, q_o + \lambda^{-r}\left(q- \frac{x_1^\lambda q}{x_1^\lambda q_o}\right) \in K^+ \quad . \qquad (6.5)$$

As $\lambda \to \infty$, since C_1 is compact , $x_1^\lambda \to \bar{x}_1$ over some subsequence; and since M is continuous and K^+ closed,

$$\bar{x}_1 M(\bar{x}_1) \leq 0 \qquad (6.6)$$

and

$$M(\bar{x}_1) - \frac{\bar{x}_1 M(\bar{x}_1)}{\bar{x}_1 q_o}\, q_o \in K^+ \quad . \qquad (6.7)$$

If $\bar{x}_1 M(\bar{x}_1) = 0$, $\bar{x}_1$ solves the complementarity problem with $q = 0$, a contradiction.   If $\bar{x}_1 M(\bar{x}_1) < 0$, then

$\alpha \equiv -\bar{x}_1 q_0 / \bar{x}_1 M(\bar{x}_1) > 0$, and from (6.7), $M(\alpha^{1/r} \bar{x}_1) + q_0 \in K^+$.

Thus $\bar{y} \equiv \alpha^{1/r} \bar{x}_1$ is another solution of the complementarity problem with $q = q_0$, a contradiction. This means for all λ sufficiently large, C_λ has no almost-complementary points, and so corollary (6.1) follows from Theorem (5.3).

Another sufficient condition is the following.

(6.8) <u>Corollary (Karamardian [14])</u> The GCP has a solution if there is a non-empty compact set $D \subset K$ such that for each $x \in K$ and not in D , there exists $z \in D$ such that $(x-z)w(x) > 0$.

Under the conditions of (6.8) it can again be shown that for λ sufficiently large, C_λ contains no almost-complementary points.

To apply Theorem (5.3), all that is actually required is the existence of a set which contains no almost-complementary points and "cuts off" zero from infinity. The following corollary results from hypothesizing that for <u>some</u> $p \in$ interior K^+ , and <u>some</u> $\lambda > 0$, $C_\lambda \equiv \{x \in K \mid xp = \lambda\}$ contains no almost-complementary points.

(6.9) <u>Corollary</u> The GCP has a solution if there is a $p \in K^+$ and $\lambda > 0$ such that for each $x \in C_\lambda$, there exists $z \in K$, $zp \leq \lambda$, such that $(x-z)w(x) > 0$.

<u>Proof</u>: We shall show that if $x \in C_\lambda$, then x is not an almost-complementary point. Suppose $x \in C_\lambda$ is almost-

complementary. Then $\tau \equiv -xw(x)/xp = -xw(x)/\lambda \geq 0$, and

$(x(x)+\tau p) \in K^+$. Suppose $y \in K$, $yp \leq \lambda$. Then

$0 \leq y(w(x)+\tau p) = yw(x) - \dfrac{xw(x)}{\lambda} yp \leq yw(x) - xw(x)$, where

the first inequality is because $(x(x)+\tau p) \in K^+$ and the
second inequality follows from $yp \leq \lambda$ and $-xw(x) \geq 0$.
Thus there can be no $z \in K$, $zp \leq \lambda$, $(x-z)w(x) > 0$. Again,
the corollary follows from Theorem (5.8).

Acknowledgments

Most of the work reported here was done as a part of my
thesis at Stanford University while I was a participant of
the Doctoral Support Plan of Bell Laboratories.

I wish to thank Professors Robert Wilson, B. Curtis
Eaves, and R. W. Cottle for much assistance in the original
effort; and Romesh Saigal and Stepan Karamardian for encour-
aging me to get it into its current form.

REFERENCES

[1]   Berge, C. "Topological Spaces," translated by E. M.
      Patterson, The Macmillan Company, New York, 1963.

[2]   Browder, F. E. "On Continuity of Fixed Points Under
      Deformations of Continuous Mappings," Summa Brasil.
      Math., Vol. 4, 1960.

[3]   Cottle, R. W. "Nonlinear Programming with Positively
      Bounded Jacobians," SIAM J. Appl. Math., Vol. 14, 1966.

[4]   Cottle, R. W., and Dantzig, G. B. "Complementary Pivot
      Theory of Mathematical Programming," Linear Algebra and
      Appl., Vol. 1, 1968, pp. 103-125.

[5]   Eaves, B. C. "On the Basic Theorem of Complementarity,"
      Journal of Optimization Theory and Applications, Vol. 7,
      No. 4, 1971.

[6]   Eaves, B. C. "Computing Kakutani Fixed Points," SIAM
      J. Applied Math., Vol. 21, No. 2, 1971, pp. 236-244.

[7]   Eaves, B. C. "Homotopies for Computation of Fixed
      Points," Math. Programming, 3(1), 1972, pp. 1-22.

[8]   Eaves, B. C., and Saigal, R. "Homotopies for Computa-
      tion of Fixed Points on Unbounded Regions," Math. Pro-
      gramming, 3(2), October 1972.

[9]   Freidenfelds, J. "Fixed Point Algorithms and Almost-
      Complementary Sets," Technical Report No. 71-17, Dept.
      of Operations Research, Stanford University, Stanford,
      California, December 1971.

[10] Habetler, G. J., and Price, A. L. "Existence Theory
      for Generalized Nonlinear Complementarity Problems,"
      Journal of Optimization Theory and Applications, Vol. 7,
      No. 4, 1971.

[11]   Hansen, T.   "On the Approximation of a Competitive
       Equilibrium," Ph.D. Dissertation, Yale University,
       New Haven, Connecticut, 1968.

[12]   Kakutani, S.   "A Generalization of Brouwer's Fixed
       Point Theorem," Duke Math. Journal, Vol. 8, 1941.

[13]   Karamardian, S.   "The Nonlinear Complementarity Prob-
       lem with Applications," Parts I and II, Journal of
       Optimization Theory and Applications, Vol. 4, No.'s 2
       and 3, 1969.

[14]   Karamardian, S.   "Generalized Complementarity Problem,"
       Journal of Optimization Theory and Applications, Vol. 8,
       No. 3, 1971, pp. 161–168.

[15]   Karamardian, S.   "An Existence Theorem for the Comple-
       mentarity Problem," Technical Report No. 159, November
       27, 1973, Department of Math. Sciences, Clemson Univer-
       sity, Clemson, South Carolina.

[16]   Kuhn, H. W.   "Simplicial Approximation of Fixed Points,"
       Proc. Nat. Acad. Sci. U.S.A., Vol. 61, 1968, 1238–1242.

[17]   Lemke, C. E.   "Bimatrix Equilibrium Points and Mathe-
       matical Programming," Management Sci., Vol. 11, 1965,
       pp. 681–689.

[18]   Lemke, C. E., and Howson, J. T., Jr. "Equilibrium
       Points of Bimatrix Games," J. Soc. Indust. Appl. Math.,
       Vol. 12, 1964, pp. 413–423.

[19]   Merrill, O. H.   "Applications and Extensions of an
       Algorithm that Computes Fixed Points of Certain Non-
       empty, Convex, Upper Semi-Continuous Point to Set Map-
       pings", Technical Report 71-7, Dept. of Industrial En-
       gineering, University of Michigan, September 1971.

[20]   Scarf, H.   "The Approximation of Fixed Points of a Con-
       tinuous Mapping," SIAM J. Appl. Math., Vol. 15, 1967,
       pp. 1328–1343.

[21]   Scarf, H.   "The Core of an N Person Game," Econometri-
       ca, Vol. 35, 1967, pp. 50–69.

A Computational Comparison of Fixed
Point Algorithms Which Use Complementary Pivoting

Richard Wilmuth

ABSTRACT

The recently developed collection of algorithms which
use the technique of complementary pivoting on a complex for
computing fixed points of functions is considered. The results of a study comparing their relative computational performance on fixed point problems from economics and nonlinear
programming are given.

1. Introduction

Let C be a subset of R^n and $f:C \to C*$ where $C*$
is the collection of all subsets of C . A point $x \in C$ is
said to be a <u>fixed point</u> of f if and only if $x \in f(x)$.
During the past few years several algorithms have emerged for
computing, in the limiting sense, fixed points of f .
These algorithms are similar in that they all operate on a
special class of complexes termed pseudomanifolds and their
convergence is based on the complementary path method of
Lemke and Howson [8].

Presently, there exist four classes of such algorithms,
each basically characterized by the type of pseudomanifold
upon which the algorithms operate. First, a brief characterization of the algorithms in each class is given. The results of a study comparing the relative computational performance of representative algorithms from each class on fixed
point problems from economics and nonlinear programming are
then presented.

2. The S^n Class Algorithms

This class of algorithms was the first to appear and
is restricted to computing fixed points of functions
$f:S^n \to S^{n*}$ on the standard n-simplex, S^n . Scarf's algorithm [10] was the first to emerge and served as a basis for
the other procedures of Kuhn [7] and later Eaves [3 and 4].

Each of these algorithms operates on essentially the same set of points or triangulations of s^n . This triangulation is the uniform grid illustrated in Figure 1 for s^2 and has come to be known as "Kuhn's triangulation."

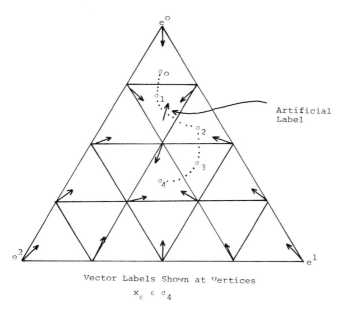

Vector Labels Shown at Vertices
$$x_\varepsilon \in \sigma_4$$

Kuhn's Triangulation of s^2

Figure 1

Roughly, these algorithms operate on a grid of "size $\varepsilon > 0$" by determining, in a finite number of steps, an exact fixed point x_ε to a continuous piecewise affine approximation $f_\varepsilon : s^n \to s^n$ obtained by affinely extending the values of f at the vertices of the grid on s^n . Under certain conditions on f , the sequence $\{x_\varepsilon\}$ can be shown to tend to fixed points of f as $\varepsilon \to 0$.

The algorithms compute each x_ϵ by generating a unique finite sequence of "adjacent" n-simplexes $\{\sigma_o, \ldots, \sigma_t\}$ on the triangulation (see Figure 1), x_ϵ then being a convex combination of the vertices of σ_t . Each vertex, v , of the grid is "labeled" with a vector, $\ell(v)$, representing the action of the function f at the vertex. At each step, a canonical tableau representing a convex combination of an artificial label with the labels of n of the vertices of σ_i is obtained. Upon pivoting into the tableau the label corresponding to the $(n+1)^{th}$ vertex of σ_i , a label corresponding to another vertex may be dropped; σ_{i+1} is defined to be that n-simplex adjacent to σ_i and "opposite" the dropped vertex. Should the artificial column be dropped, the procedure is terminated with the desired simplex, σ_t . The complementary path technique of Lemke and Howson is used to show that no simplex is revisited and hence finite termination occurs.

The major disadvantage of these algorithms is that in order to generate x_δ for some $\delta < \epsilon$, the entire algorithm must be repeated. This is somewhat of a problem since the average number of iterations to find each x_ϵ appears to be a linear function of $1/\epsilon$.

If $f:S^n \to S^n$ is a point-to-point function, Scarf [10] and Kuhn [7] replace the vector labels with integer labels. The tableau is replaced by the correspondences between a set of distinct integers (one of which is artificial) and the vertices. Pivoting then becomes simply a matter of maintaining the proper correspondence between the integer labels and the vertices. The procedure is terminated when the artificial label is dropped.

Using the integer labels, each step is much less complicated and obviously faster. However, since less information is provided in an integer than a vector, more steps may be required. Both integer and vector labels require one evaluation of f for each pivot. If the function evaluations are time-consuming, then the number of steps required becomes more important than the complexity of the pivot operation.

Eaves' algorithm [4] was chosen to represent this class of algorithms in the computational study. The special case of the algorithm used (in terms of the artificial label) is essentially the same as Scarf's algorithm with vector labels.

3.   The $S^n \times [0,\infty)$ Class Algorithms

This class consists solely of the algorithm of Eaves [5] which is also restricted to computing fixed points of functions $f:S^n \to S^{n^*}$ on the standard n-simplex. This was the first algorithm to continuously calculate approximate fixed points of f as the grid size becomes finer and hence overcame the problem of requiring the algorithm to be rerun to achieve a finer grid.

Eaves uses an auxilary dimension and imposes his grid structure on the (n+1)-dimensional cylinder $S^n \times [0,\infty)$.

The triangulation of $S^n \times [0,\infty)$ generally used with Eaves' algorithm is illustrated in Figure 2 for n = 1 and has the

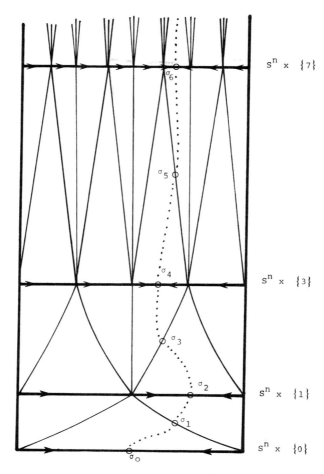

Eaves' Triangulation of $S^n \times [0,\infty)$

Figure 2

property that all vertices of the triangulation are contained
in the cross-sections $S^n \times \{k\}$ for $k = 0,1,3,7,15,\ldots,$
i.e., $k = 2^j-1$, $j = 0,1,2,3,\ldots,$ and the set of all faces in
each cross-section constitute Kuhn's triangulation of size

$1/(k+1)$.   Hence, all the   (n+1)-dimensional simplexes in $S^n \times [\frac{k-1}{2}, k]$ are of "height" $\frac{k+1}{2}$ in the auxilary dimension and movement up the cylinder by one cross-sectional level increases the fineness of the grid by a factor of 2   .

The vertices in the interior of the cylinder are labeled (vector or integer) as per the action of the function on their projection onto S^n .   All points on the boundary are labeled to reflect an artificial action to a constant point in the interior of S^n .   Starring with $\sigma_o = S^n \times \{0\}$, one may define an <u>infinite</u> sequence $\{\sigma_o, \sigma_1, \ldots, \}$ of adjacent n-simplexes by pivoting (adding and dropping vertices) in a manner analogous to that of the previous algorithm.   The tableau (assuming vector labels) always yields a convex combination of the labels and hence the projection of σ_i onto S^n is an approximate fixed point of   f   as long as none of the vertices of σ_i are on the boundary of the cylinder.

The artificial labeling provides a means for starting the algorithm and keeps the sequence $\{\sigma_o, \sigma_1, \ldots\}$ <u>within</u> the cylinder.   Since there is a finite number of simplexes below each level of the cylinder and no simplex is ever revisited, the algorithm will be assured of reaching any grid fineness within a finite number of iterations.

The vector and integer labels here have the same performance trade-off between individual pivoting and the number of pivots required as discussed with the previous algorithms. Although this algorithm reduces grid fineness automatically, the triangulation is considerably more complex and rigidly dictates that this reduction be accomplished by factors of 2.

4. The $R^n \times [0,\infty)$ Class Algorithms

The algorithm of Eaves and Saigal [6] is the single member of this class and is an extension of Eaves' algorithm on $S^n \times [0,\infty)$ to handle functions $f:R^n \to R^{n*}$. This extension allows one to avoid the messy, and sometimes virtually impossible, procedure of embedding the domain of f in the standard n-simplex. As well, the technique used permits one to restart the procedure at any point with any grid size and thereby gives increased flexibility in initial and intermediate grid sizes.

Eaves and Saigal extended Eaves' original triangulation of $S^n \times [0,\infty)$ by first triangulating $R^n \times \{0\}$ with Kuhn's triangulation of some initial grid size, δ . A "scaled" copy of Eaves' triangulation of $S^n \times [0,\infty)$ is then placed on each simplex of Kuhn's triangulation in a manner that facilitates movement between cylinders at all levels above the base. This is illustrated in Figure 3 for $n = 1$.

Eaves and Saigal then used the convex cone, shown in Figure 3 as a vehicle to provide an initial starting simplex for the algorithm and to force movement to finer and finer grid sizes. Below the cone, vertices of the grid are artificially labeled (always with vectors) to force movement "inward" towards the cone. Within the cone, vertices are labeled (always with vectors) to reflect the true action of the function at the vertex projected onto the base $R^n \times \{0\}$. This technique provides a starting simplex at the vertex of the cone. The artificial labeling keeps the algorithm operating within the cone. Since there is a finite

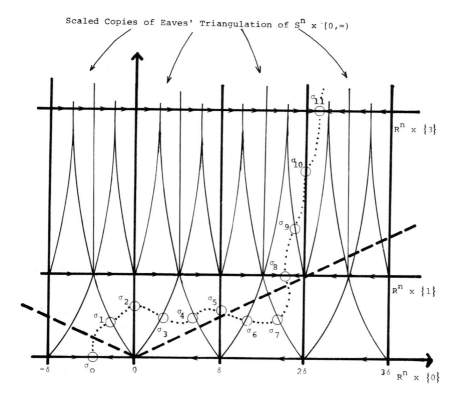

Scaled Copies of Eaves' Triangulation of $S^n \times [0,\infty)$

Eaves' and Saigal's Triangulation of $R^n \times [0,\infty)$

Figure 3

number of simplexes below any given level of the cone, and
no simplex is ever revisited, the sequence $\{\sigma_o, \sigma_1, \dots\}$
will therefore reach that given level of grid fineness in a
finite number of iterations.

5. The R^n x [0,1] Class Algorithms

The algorithm of Merrill [9] constitutes the single element of this class. As with the Eaves/Saigal algorithm, this algorithm handles functions on R^n and thereby also avoids the procedure of embedding the domain of f in the standard n-simplex.

This algorithm impresses a simple triangulation of given grid fineness δ on the region R^n x [0,1] as shown in Figure 4 for n = 1 . On the R^n x {1} cross-section, the vertices are labeled to reflect the action of the function f on their projections onto R^n . The vertices within R^n x {0} are artificially labeled in a manner which provides an initial simplex σ_0 in R^n x {0} containing the desired starting point $x^o \in R^n$. This artificial labeling is also chosen to guarantee that the algorithm will not visit any n-simplex in R^n x {0} other than the initial one. Under certain conditions on f , a finite sequence $\{\sigma_o,\ldots,\sigma_t\}$ of n-simplexes will be generated with

$\sigma_t \subset R^n$ x {1} and the projection of σ_t approximating a fixed point of f .

The grid may then be refined and the algorithm rerun with $x^1 \in$ conv (σ_t) as the new starting point. Successive grid refinements and rerunning of the algorithm will produce a sequence of approximate fixed points which converge to fixed points of f . With this algorithm the grid refinement may be of any amount and even different at each stage.

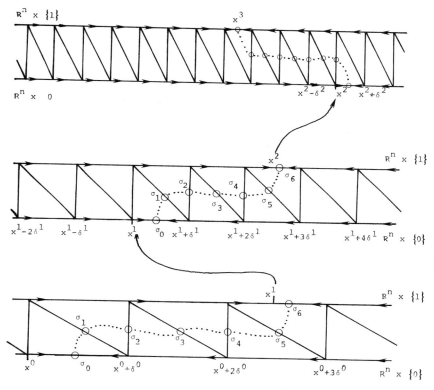

Merrill's Triangulation of $R^n \times [0,1]$

Figure 4

To achieve this flexibility, with each grid refinement the artificial labeling must be reintroduced, possibly throwing away useful information and hence increasing the probability for extra iterations. However, this loss in performance may be compensated in the simplicity of the complex used.

6. Algorithm Comparison

Each iteration in all of the algorithms considered here consists of i) a single function evaluation, ii) a tableau pivot, and iii) the determination of an adjacent simplex. The measure used to compare the relative performance of the algorithms on each test problem is the level of grid fineness attained after a given number of iterations. This measure provides an implementation independent means of comparing the algorithms.

The algorithm of Eaves on S^n was taken as representative of all three algorithms in the S^n class. As implemented, it is essentially the same as Scarf's algorithm with vector labels. The three algorithms constituting the other three classes were all implemented for the study. In the case of Eaves' algorithm on $S^n \times [0,\infty)$ both the integer and vector labeling methods were implemented to give a performance comparison of the two labeling techniques. The algorithms were all coded in the programming language APL/360 [1] and these codes may be found in Wilmuth [11].

The algorithms were run on eleven test problems from economics and nonlinear programming. The first three problems were to compute approximate fixed points for 3 cases of a continuous function $f:S^n \to S^n$ (with $n = 4$, 7 and 9) whose fixed points are "equilibrium price vectors" in the pure exchange model given in Scarf [10]. The remaining eight problems were to compute approximate fixed points for eight point-to-set maps $f:R^n \to R^{n*}$ whose fixed points are stationary points of nonlinear programming problems. The definition of f in terms of the standard nonlinear program min $\{\theta(x) \mid g(x) \leq 0\}$ where $\theta: R^n \to R$ and $g: R^n \to R^n$ is that given in Merrill [9]. The eight problems are the well-known Colville test problems and are completely described in Colville [2].

The results for Scarf's problems and the Colville problems are given in Figures 5-17. The graphs depict the grid size attained after a given number of iterations. The following notation is used to indicate the input parameters used in executing each of the algorithms: Δ denotes the initial coordinate grid lengths, x^o is the starting point used, α is the "slope" of the cone in the Eaves/Saigal algorithm, and γ is the grid refinement factor used in Merrill's algorithm.

The first three graphs (figures 5, 6 and 7) give the results on Scarf's problems using the five different algorithms. Eaves' S^n class algorithm performed so badly compared to the others that it could not be plotted in two of the cases without distorting the other results. Intuitively, since doubling the accuracy in Eaves' S^n algorithm requires roughly doubling the number of iterations previously taken, one readily sees that this algorithm is not comparable in performance with the other four.

Eaves' $S^n \times [0, \infty)$ algorithm using integer labels also did not perform too well. The difficulty with this algorithm appeared to be that the integer labeling did not identify a "good" path to the fixed point. After some time, the algorithm began converging to the correct fixed point but only very slowly and using a very small grid size.

The relative performance of the remaining three algorithms can be seen easily from the graphs. These algorithms were terminated when $|f(x) - x| < 0.01$.

Since the Colville problems give rise to point-to-set functions on R^n , only the Eaves/Saigal R^n x $[0,\infty)$ algorithm and Merrill's R^n x $[0,1]$ algorithm were tested on these problems. Of course, in the limit, the Eaves/Saigal R^n x $[0,\infty)$ algorithm should perform the same as Eaves' S^n x $[0,\infty)$ algorithm.

Figures 8-17 give the results for the Colville problems. Convergence to a stationary point was achieved for all problems except for problem 2 and in each case all algorithms converged to the same point. Convergence for problem 2 was too slow to be practically carried out using an APL implementation of the algorithms. That convergence did occur with Merrill's algorithm for problem 2 was verified by communication with Dr. Merrill who observed convergence to a grid size of 0.086 after 84,394 iterations using a grid refinement factor of 0.58. A graph of the results of convergence for the function $f(x) = (x_1^3,\ldots,x_n^3)$ has been substituted for problem 2 to show that the slow convergence is due to the complexity of the problem rather than the dimension (n = 15).

Observe that the Eaves/Saigal algorithm generally takes significantly fewer iterations to reach a given level of grid refinement than Merrill's algorithm does in problems 1, 3, 6, 7, 8 and $f(x) = x^3$; whereas, the converse is true for problems 4 and 5.

The table in Figure 18 gives the average processor time per iteration for each algorithm on each problem as run on a System 370/85. Note that Merrill's algorithm takes about one-half the processor time per iteration of the Eaves/Saigal algorithm. Hence, in an APL implementation these two algorithms appear to perform about the same. The difference between the two algorithms can be accounted for in the computing of the complicated pivot rules of the Eaves/Saigal complex versus the simple complex used by Merrill. It is conjectured that in a PL/I or Assembly language implementation of these algorithms, the Eaves/Saigal pivot rules could be programmed to take about the same processor time per iteration as Merrill's.

Finally, a comment should be made concerning integer labeling techniques. The studies reported here showed integer labeling to require several times the number of iterations to reach a given grid fineness as compared to vector labeling techniques. For complicated functions in which the function evaluation consumes a large amount of computation time it is important to minimize the number of iterations and in this case vector labeling appears to be the better approach. If, however, function evaluations are simple and the vector tableau pivot consumes most of the processor time per iteration, then the large number of iterations which appear to be required for integer labeling might be compensated for by the significantly reduced processor time per iteration one might obtain with the more simple integer pivot operation. In the later case, integer labeling mathods could prove to be superior.

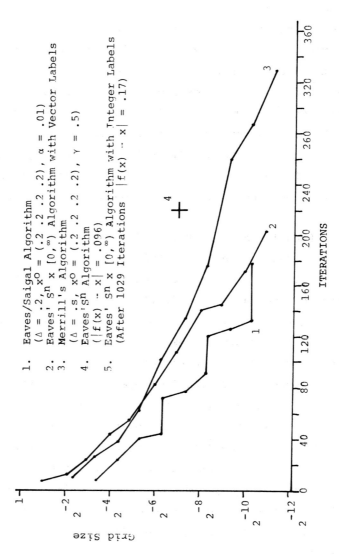

SCARF'S PROBLEM 1

(n=4)

1. Eaves/Saigal Algorithm
   (Δ = .2, x^0 = (.2 .2 .2 .2), α = .01)
2. Eaves' S^n x [0,∞) Algorithm with Vector Labels
3. Merrill's Algorithm
   (Δ = .s, x^0 = (.2 .2 .2 .2), γ = .5)
4. Eaves' S^n Algorithm
   ($|f(x) \cdot x|$ = .096)
5. Eaves' S^n x [0,∞) Algorithm with Integer Labels
   (After 1029 Iterations $|f(x) \cdot x|$ = .17)

ITERATIONS

Grid Size

Figure 5

264

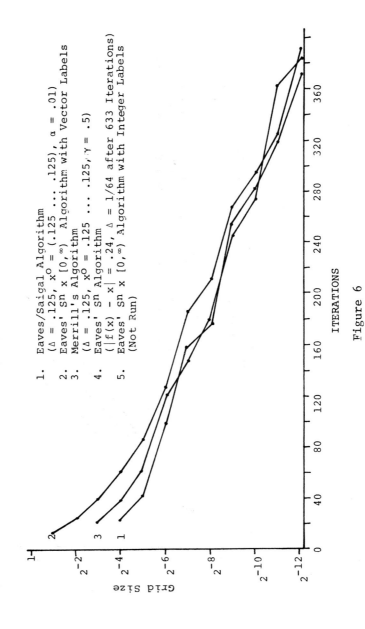

SCARF'S PROBLEM 2

(n=7)

1. Eaves/Saigal Algorithm
   (Δ = .125, x^0 = (.125 .... .125), α = .01)
2. Eaves' S^n x $[0,\infty)$ Algorithm with Vector Labels
3. Merrill's Algorithm
   (Δ = .125, x^0 = .125 .... .125, γ = .5)
4. Eaves' S^n Algorithm
   ($|f(x) - x|$ = .24, Δ = 1/64 after 633 Iterations)
5. Eaves' S^n x $[0,\infty)$ Algorithm with Integer Labels
   (Not Run)

ITERATIONS

Figure 6

265

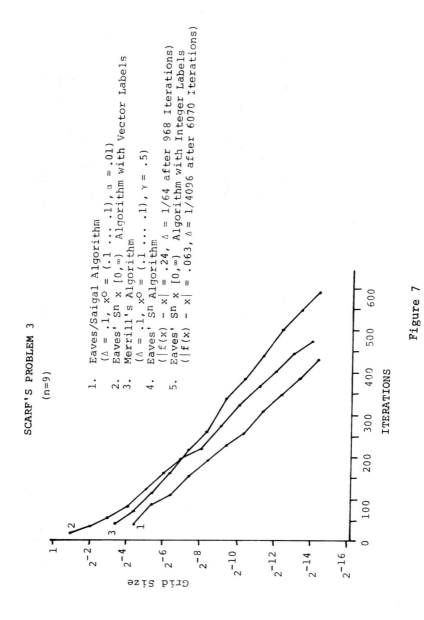

SCARF'S PROBLEM 3

(n=9)

1. Eaves/Saigal Algorithm
   (Δ = .1, x^0 = (.1 ... .1), α = .01)
2. Eaves' S^n x [0,∞) Algorithm with Vector Labels
3. Merrill's Algorithm
   (Δ = .1, x^0 = (.1 ... .1), γ = .5)
4. Eaves' S^n Algorithm
   ($|f(x) - x|$ = .24, Δ = 1/64 after 968 Iterations)
5. Eaves' S^n x [0,∞) Algorithm with Integer Labels
   ($|f(x) - x|$ = .063, Δ = 1/4096 after 6070 iterations)

ITERATIONS

Grid Size

Figure 7

266

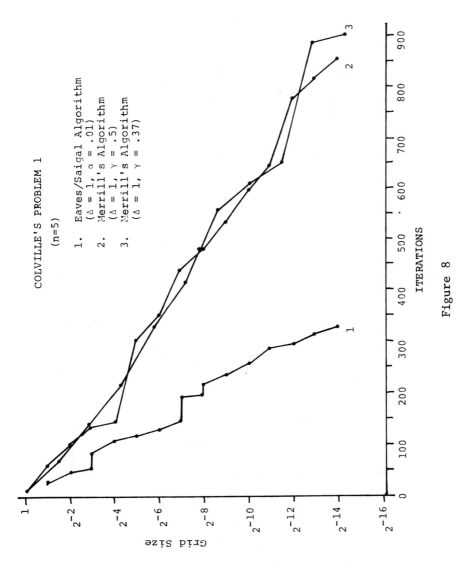

COLVILLE'S PROBLEM 1

(n=5)

1. Eaves/Saigal Algorithm
   (Δ = 1, α = .01)
2. Merrill's Algorithm
   (Δ = 1, γ = .5)
3. Merrill's Algorithm
   (Δ = 1, γ = .37)

Figure 8

267

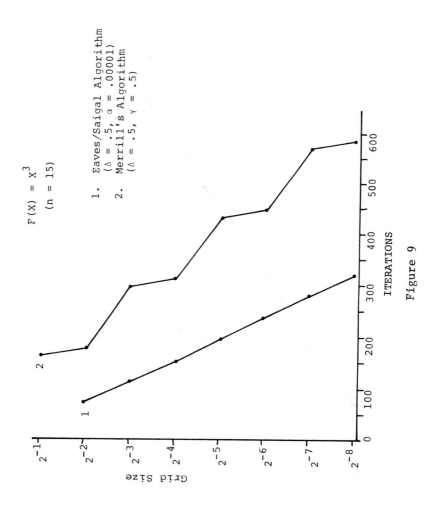

$F(X) = X^3$

$(n = 15)$

1. Eaves/Saigal Algorithm
   ($\Delta = .5$, $\alpha = .00001$)
2. Merrill's Algorithm
   ($\Delta = .5$, $\gamma = .5$)

ITERATIONS

Grid Size

Figure 9

COLVILLE'S PROBLEM 3 NONFEASIBLE STARTING POINT

(n = 5)

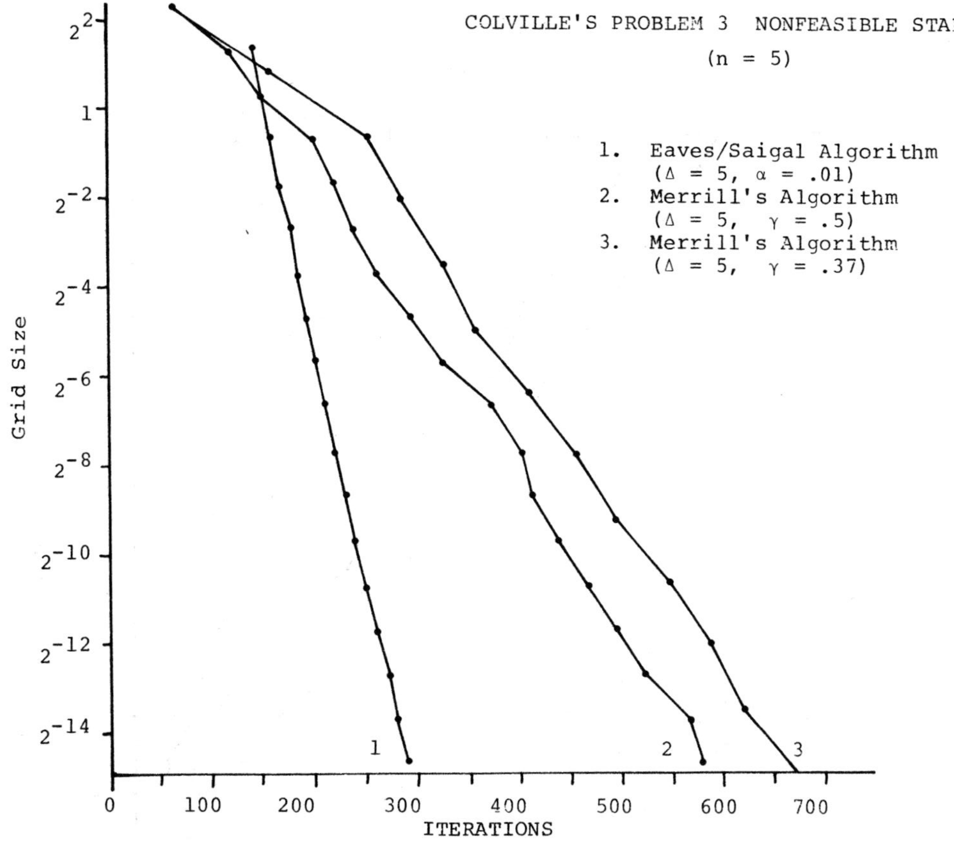
1. Eaves/Saigal Algorithm (Δ = 5, α = .01)
2. Merrill's Algorithm (Δ = 5, γ = .5)
3. Merrill's Algorithm (Δ = 5, γ = .37)
Grid Size
2^{2}
1
2^{-2}
2^{-4}
2^{-6}
2^{-8}
2^{-10}
2^{-12}
2^{-14}
0
100
200
300
400
500
600
700
ITERATIONS
1
2
3

Figure 10

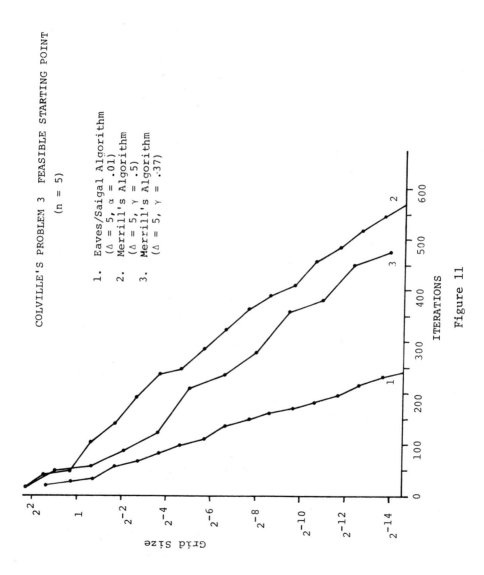

COLVILLE'S PROBLEM 3   FEASIBLE STARTING POINT

(n = 5)

1. Eaves/Saigal Algorithm
   (Δ = 5, α = .01)
2. Merrill's Algorithm
   (Δ = 5, γ = .5)
3. Merrill's Algorithm
   (Δ = 5, γ = .37)

Grid Size

ITERATIONS

Figure 11

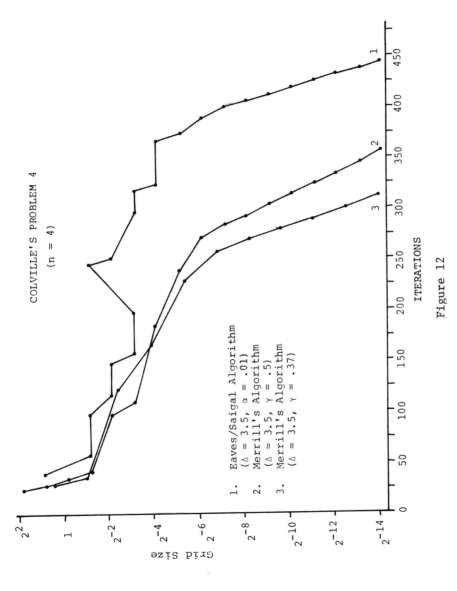

COLVILLE'S PROBLEM 4

(n = 4)

1. Eaves/Saigal Algorithm
   (Δ = 3.5, α = .01)
2. Merrill's Algorithm
   (Δ = 3.5, γ = .5)
3. Merrill's Algorithm
   (Δ = 3.5, γ = .37)

ITERATIONS

Grid Size

Figure 12

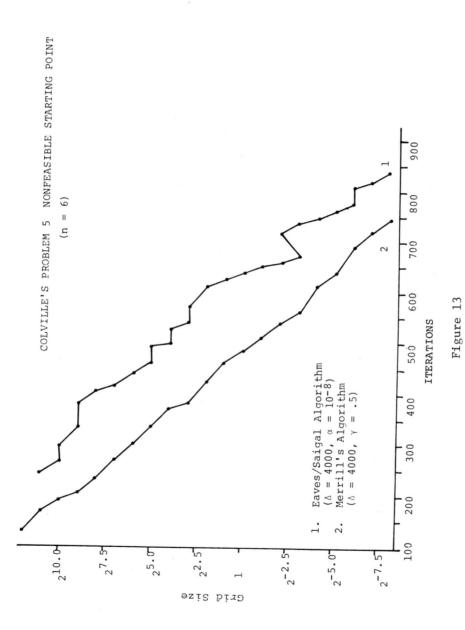

COLVILLE'S PROBLEM 5   NONFEASIBLE STARTING POINT

(n = 6)

1. Eaves/Saigal Algorithm
   (Δ = 4000, α = 10^{-8})
2. Merrill's Algorithm
   (Δ = 4000, γ = .5)

Grid Size

ITERATIONS

Figure 13

272

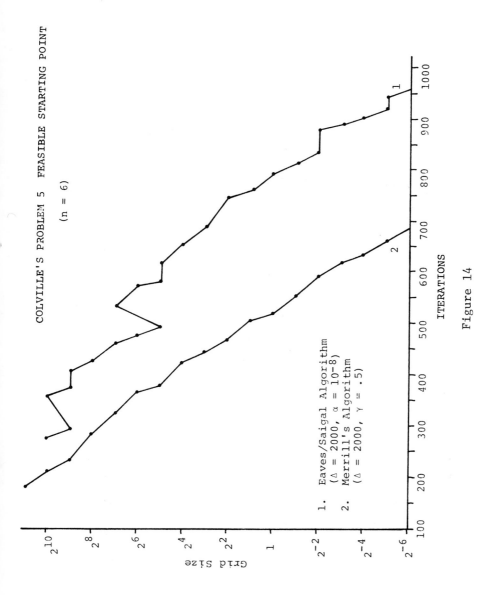

COLVILLE'S PROBLEM 5 FEASIBLE STARTING POINT

(n = 6)

1. Eaves/Saigal Algorithm
    ($\Delta = 2000$, $\alpha = 10^{-8}$)
2. Merrill's Algorithm
    ($\Delta = 2000$, $\gamma = .5$)

Grid Size

ITERATIONS

Figure 14

273

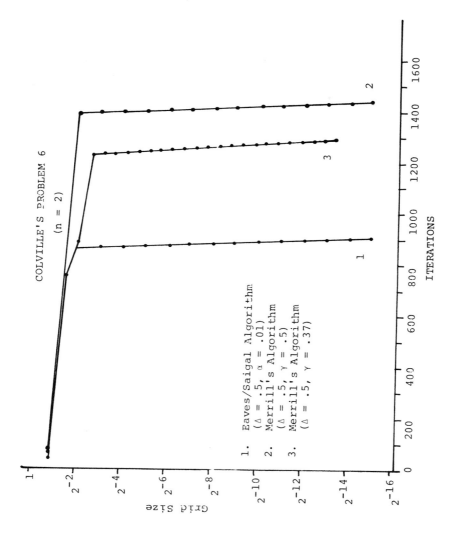

COLVILLE'S PROBLEM 6

(n = 2)

1. Eaves/Saigal Algorithm
   (Δ = .5, α = .01)
2. Merrill's Algorithm
   (Δ = .5, γ = .5)
3. Merrill's Algorithm
   (Δ = .5, γ = .37)

Grid Size

ITERATIONS

Figure 15

274

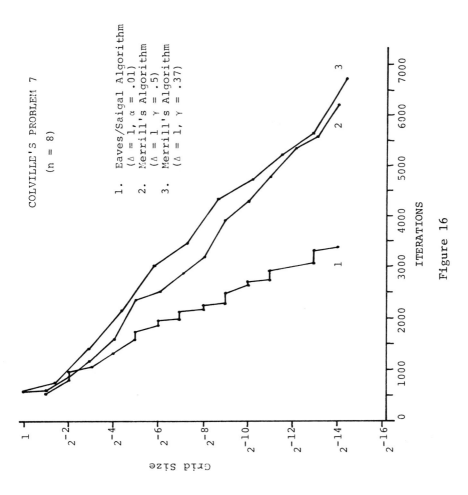

COLVILLE'S PROBLEM 7

(n = 8)

1. Eaves/Saigal Algorithm
   (Δ = 1, α = .01)
2. Merrill's Algorithm
   (Δ = 1, γ = .5)
3. Merrill's Algorithm
   (Δ = 1, γ = .37)

Grid Size

ITERATIONS

Figure 16

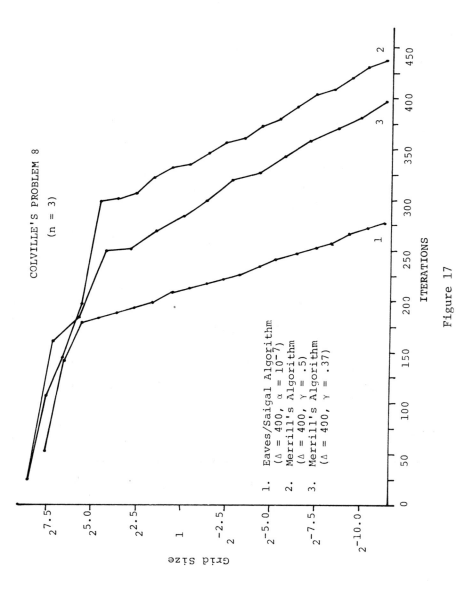

COLVILLE'S PROBLEM 8

(n = 3)

1. Eaves/Saigal Algorithm
   (Δ = 400, α = 10^{-7})
2. Merrill's Algorithm
   (Δ = 400, γ = .5)
3. Merrill's Algorithm
   (Δ = 400, γ = .37)

Grid Size

ITERATIONS

Figure 17

276

Problem	Algorithm	Ave. Seconds per Iteration
Scarf 1	Eaves/Saigal Merrill's Eaves' Vector* Eaves' S^n Eaves Integer	.054 .024 .039 .023 .028
Scarf 2	Eaves/Saigal Merrill's Eaves' Vector Eaves' S^n Eaves' Integer	.060 .034 .048 .028 ----
Scarf 3	Eaves/Saigal Merrill's Eaves' Vector Eaves' S^n Eaves' Integer	.071 .041 .055 .036 .043
Colville 1	Eaves/Saigal Merrill's	.041 .019
$f(x) = x^3$	Eaves/Saigal Merrill's	.073 .036
Colville 3	Eaves/Saigal Merrill's	.047 .023
Colville 4	Eaves/Saigal Merrill's	.044 .023
Colville 5	Eaves/Saigal Merrill's	.110 .060
Colville 6	Eaves/Saigal Merrill's	.073 .044
Colville 7	Eaves/Saigal Merrill's	.066 .033
Colville 8	Eaves/Saigal Merrill's	.133 .079

*Eaves' $S^n x [0, \infty)$ Algorithm with Vector Labels

Processor Time Comparisons

Figure 18

7. Acknowledgements

This research was carried out as part of my Doctoral Dissertation in the Department of Operations Research at Stanford University. I would like to take this opportunity to thank my advisor, Professor B. Curtis Eaves, for his excellent advice and support.

REFERENCES

[1] APL/360 User's Manual, IBM Technical Publications
     Dept., White Plains, New York, March 1970.

[2] Colville, A. R. "A Comparitive Study on Nonlinear Pro-
     gramming Codes," IBM New York Scientific Center Report
     No. 320-2949, June 1968.

[3] Eaves, B. C. "An Odd Theorem," Proceedings of the
     American Mathematical Society, 26, 3, 1970, 509-513.

[4] Eaves, B. C. "Computing Kakutani Fixed Points," SIAM
     Journal of Applied Mathematics 21, 2, 1971, 236-244.

[5] Eaves, B. C. "Homotopies for the Computation of Fixed
     Points," Mathematical Programming 3, 1972, 1-22.

[6] Eaves, B. C., and Saigal, R. "Homotopies for the Com-
     putation of Fixed Points on Unbounded Regions," Mathe-
     matical Programming 3, 1972, 225-237.

[7] Kuhn, H. W. "Simplicial Approximation of Fixed Points,"
     Proceedings of the National Academy of Sciences 61,
     1968, 1238-1242.

[8] Lemke, C. E., and Howson, J. T. "Equilibrium Points of
     Bimatrix Games," Journal of the Society of Industrial
     Applied Mathematics 12, 2, 1964, 413-423.

[9] Merrill, O. H. "Applications and Extensions of an
     Algorithm that Computes Fixed Points of Certain Non-
     Empty Convex Upper Semicontinuous Point-to-Set Map-
     pings," Department of Industrial Engineering, Univer-
     sity of Michigan, Technical Report No. 71-7, Sept. 1971.

[10] Scarf, H. "The Approximation of Fixed Points of a Con-
     tinuous Mapping," SIAM Journal of Applied Mathematics
     15, 5, 1967, 1328-1343.

[11]   Wilmuth, R. J.   "The Computation of Fixed Points,"
       Doctoral Dissertation, Department of Operations Re-
       search, Stanford University, July 1973.

Computational Experience With Large-Scale
Linear Complementarity Problems

Richard W. Cottle

ABSTRACT

This paper contains a brief summary of some computational experience acquired by the Systems Optimization Laboratory at Stanford University on linear complementarity problems of intermediate to large size.

1.   Introduction

Although these proceedings are concerned with computing
fixed points and the applications of such calculations, there
is a two-fold rationale for the inclusion of an article on
solving the linear complementarity problem. First, there is
the simple historical fact that H. Scarf's important paper
[39] on the approximation of fixed points was directly influenced by C. E. Lemke's equally important paper [30] on solving what is now called the linear complementarity problem.
Second, there is an equivalence between the complementarity
problem (linear or nonlinear) and the Brouwer fixed point
problem over the nonnegative orthant in R^n . This equivalence serves as our point of departure.

Given a mapping $F:R^n \rightarrow R^n$, the system

$$x \geq 0 \quad , \quad F(x) \geq 0 \quad , \quad \langle x, F(x) \rangle = 0 \qquad (1.1)$$

specifies a __complementarity__ __problem__ the main aim of which is
to obtain a solution if one exists. When F is an affine
transformation, say $F(x) = q + Mx$, then (1.1) is called a
__linear__ __complementarity__ __problem__ and is denoted by the pair
(q,M) . Now if H is a mapping defined by

$$h_i(x) = \begin{cases} x_i - f_i(x) & \text{if } x_i - f_i(x) > 0 \\ \\ 0 & \text{otherwise} \end{cases} \qquad i=1,\ldots,n \qquad (1.2)$$

then $H:R_+^n \to R_+^n$, and it is easily shown that a fixed point

of H solves (1.1). Conversely, if H is an arbitrary

self-mapping of R_+^n , it follows that any solution of (1.1)

with F defined through

$$F(x) = x - H(x) \qquad\qquad (1.3)$$

also solves the Brouwer fixed point problem associated with

H and R_+^n . See [28], [21], [37].

   Interesting though it might be to do so, our purpose
here is not to compare the theoretical and computational aspects of these kindred subjects, but rather to give a brief
and somewhat preliminary report on a few types of linear complementarity problems solved under the auspices of the Systems Optimization Laboratory at Stanford University. All the
computation was done on the IBM 360/91 using FORTRAN H with
Opt = 2.

   The problems reported on here range from n = 50 with
M dense to n = 4095 with M sparse and specially
structured. The computational work employs double-precision
codes for Lemke's almost-complementarity pivoting algorithm,
Modified Point S.O.R., and Modified Block S.O.R. In several
cases, the sources for the data were "real-world" models in
engineering or economics. The descriptions of these models
given here are admittedly very sketchy. However, the interested reader will find the missing details in the references
cited along the way.

2. The Linear Complementarity Problem

To fix notation, we define the linear complementarity problem (q,M) as that of solving the system

$$y = q + Mx \quad , \quad x \geq 0 \quad , \quad y \geq 0 \quad , \quad \langle x,y \rangle = 0 \qquad (2.1)$$

The components of x and y having the same index are called complements of each other. The conditions of the problem imply that any solution must satisfy $x_i y_i = 0$ for

i = 1,...,n where n is the order of M .

In general, neither existence nor uniqueness of a solution to (q,M) is guaranteed, though of course there are some fairly common circumstances in which they are. In this connection, we mention for future reference the cone K(M) consisting of all those vectors q for which the linear complementarity problem (q,M) has at least one solution. In

any case, $K(M) \supset R_+^n$. Indeed, K(M) is always the union of complementary cones.

$$\text{pos } B = \left\{ q \,\middle|\, q = \sum_{i=1}^{n} \lambda_i B_i, \ \lambda_i \geq 0, \ i = 1,\ldots,n \right\}$$

where for each i, $B_i \in \{I_i, -M_i\}$, and $B = [B_1,\ldots,B_n]$.

When M has positive principal minors, $K(M) = R^n$, and, in

fact, the complementary cones collectively partition R^n

(see [38]). When M is (element-wise) positive, $K(M) = R^n$ but the complementary cones may have interior points q in common, and this means such problems have multiple solutions (see [32]). When M is positive semi-definite, $K(M) =$ pos$(I, -M)$. More generally, this statement is true when M is "copositive-plus," that is, when (i) $x^T Mx \geq 0$ for all $x \geq 0$, and (ii) $(M + M^T) x = 0$ if $x \geq 0$ and $x^T Mx = 0$ (see [30], [10]). While these remarks represent only a fragment of what is known about the existence of solutions, they are enough to cover the cases under discussion here. For some up-to-date work on existence questions, see [23] and [41].

3. Lemke's Algorithm and the NULEMKE Code

The algorithm of Lemke already cited is a pivoting scheme applied to the auxiliary system

$$y = q + Mx + e\zeta \; , \; x \geq 0 \; , \; y \geq 0 \; , \; \zeta \geq 0 \; , \; \langle x,y \rangle = 0 \qquad (3.1)$$

in which e is a strictly positive n-vector. A solution of (3.1) with $\zeta = 0$ furnishes a solution of (2.1), that is, (q,M) . In essence, the method works as follows:

1° Set $x = 0$ and ζ at the least positive[*] number $\bar{\zeta}$ which will make $y \geq 0$. Make an exchange between ζ and the[**] (dependent) y-variable which vanishes when $\zeta = \bar{\zeta}$. (This y-variable is called the blocking variable.)

---

[*]We may assume $q \notin R_+^n$.

[**]Uniqueness is assumed here for simplicity.

$2°$ Increase the complement of the blocking variable as much
as possible subject to the condition that all the dependent variables remain nonnegative. (This increasing
variable is called the <u>driving variable</u>.) If no dependent variable decreases as the driving variable increases,
then terminate. Otherwise there is a blocking variable;
exchange the blocking variable and the driving variable.

$3°$ If the blocking variable is ζ , terminate. Otherwise,
return to Step 2 using the complement of the blocking
variable as the driving variable.

The procedure is finite. If termination occurs in
Step 3, a solution is found. If termination occurs in Step
2, a solution may or may not exist, depending on the nature
of the matrix M being used. However, for the types of
problems most frequently encountered, termination in Step 2
signifies the inconsistency of the inequality constraints,
i.e., that $q \notin pos(I, -M)$.

The calculations required in executing Lemke's algorithm are very much like those of the simplex method for linear programming [15] -- with, of course, special rules gov-
erning the selection of a column to enter the basis. This
being so, only two column vectors enter into the computation
by which the blocking variable (if any) is determined. This
requires updated versions of the relevant columns; these can
be constructed from original data and the inverse of the current basis. The code named NULEMKE is based on these ideas.
It couples the updating scheme via product form of the inverse from J. A. Tomlin's LPM1 code for linear programming
with an implementation of Lemke's algorithm written by D.
Cone. The NULEMKE code has been written and tested by A.
Djang, R. E. Doherty, and C. Engles.

4. Dense Medium-Size Problems

Over the years, one of the fruitful areas of application for linear complementarity has been structural mechanics.
In many instances, the problems under consideration are
equivalent to convex quadratic programs. This has allowed
investigators to use quadratic programming codes to obtain
numerical solutions for their problems and also to use the
optimization aspect of quadratic programming and its related
duality theory to establish extremal properties of the structures being analyzed.

Examples of this approach are conspicuous in the work
of G. Maier and his colleagues. (See [17], [18].) Professor
Maier has kindly provided data pertaining to a reinforced
concrete beam problem described in [18, p. 318]. (See also
[31].) In this problem, the matrix of the linear complementarity problem denoted M^* is dense, symmetric, positive
definite and of order 130. The vector $q^* \in R^{130}$ is the one
originally provided. Due to a sign mistake, we solved both
(q^*, M^*) and $(-q^*, M^*)$. As a further experiment we solved
$(-M^*e, M^*)$ because we knew the answer and a lower bound on
the number of pivots that would be required. The results of
this set of experiments are summarized in Table 1 below.

Problem	Iterations	Execution Time
(q^*, M^*)	20	0.14 seconds
$(-q^*, M^*)$	105	2.21 seconds
$(-M^*e, M^*)$	130	3.22 seconds

TABLE 1

It is difficult to make accurate comparisons of computations done on different machines and in different environments. Nontheless it is our impression that the NULEMKE execution time for solving (q^*, M^*) stated in Table 1 compares rather favorably with the 40 seconds using a modified gradient method on a UNIVAC 1106 reported by DeDonato and Franchi [17, p. 22].

In an effort to take the effects of a time-staged loading history into account, C. Polizzotto [34] has formulated a multistage linear complementarity problem (q, M) in which

$$
q = \begin{pmatrix} q_1 \\ q_2 \\ \cdot \\ \cdot \\ \cdot \\ q_m \end{pmatrix}
\qquad
M = \begin{bmatrix} A_1 & & & \\ A_1 & A_2 & & \\ \cdot & \cdot & \cdot & \\ \cdot & \cdot & \cdot & \\ \cdot & \cdot & \cdot & \\ A_1 & A_2 & \cdots & A_m \end{bmatrix}
$$

In this model, the q_i are vectors in R^p and the A_i are p x p matrices, $i = 1, \ldots, m$. The special structure of the matrix M suggests a "natural, back-substitution" scheme for solving the problem. One should begin by noting that if $\bar{z}$ solves (q, M), and $\bar{z}$ is decomposed in the same way as q , then (for $i = 1, \ldots, m$) $\bar{z}_i$ must solve $(q_i + \sum_{j<i} A_j \bar{z}_j, A_i)$. The step-by-step solution procedure this suggests is obvious. Moreover, using it would not mean m separate trips to the computation center, one for each stage. Solving (q, M) in this way simply reduces the demands on high-speed memory and

data manipulation in what could potentially be a very large
linear complementarity problem.

Although this strategy looks attractive, a caveat must
be stated. In some relatively tame situations, the non-
uniqueness of solutions in one stage can produce ambiguities
at subsequent stages. For instance, in the very simple case
where $m = p = 2$ and A_1 is positive semi-definite (but not
symmetric), we can encounter the possibility that (q_1, A_1)
has solutions $\bar{z}_1'$ and $\bar{z}_1''$ such that $(q_2 + A_1\bar{z}_1', A_2)$ is
solvable and $(q_2 + A_1\bar{z}_1'', A_2)$ is $\underline{not}$ solvable. As a matter
of fact, if A is positive semi-definite and $K(A)$ is the
cone defined in Section 2, then there is a ray of solutions
to (q, A) emanating from any one of its solutions $\bar{z}$ if and
only if q belongs to the boundary of $K(A)$. See [8].
Now, the mere existence of multiple solutions for the linear
complementarity problem (q_1, A_1) does not ipso facto cause
trouble in the succeeding problem $(q_2 + A_1\bar{z}_1, A_2)$. For instance, if $K(A_2) = R^P$, the value of $q_2 + A_1\bar{z}_1$ is of no
great consequence insofar as the solution of $(q_2 + A_1\bar{z}_1, A_2)$
is concerned. But if $\bar{z}_1'$ and $\bar{z}_1''$ are distinct solutions of
(q_1, A_1) and $A_1z_1' \neq A_1z_1''$, then for certain kinds of matrices, A_2 , it can happen that $q_2 + A_1z_1' \in K(A_2)$ and
$q_2 + A_1z_1'' \notin K(A_2)$. Fortunately, this cannot be the case
when A_1 is symmetric and positive semi-definite. See [27],
[7].

An experiment was performed to assess what advantage
there might be in using the back-substitution scheme

suggested above in place of the more routine solution of the entire problem all at once by the NULEMKE code. The size parameters for the problems solved were $m = 10$, $p = 20$ so that M had order $n = 200$. While the back-substitution scheme obviously applies to much larger problems, we decided to keep the size within comfortable reach of the NULEMKE code. Furthermore, since the point was to compare efficiency, it was decided to generate necessarily solvable problems; in particular, the A_i $(i = 1,...,m)$ were first made to be positive definite and symmetric. The results are recorded in Table 2.

Initializer	Pivots (NULEMKE)	Time (NULEMKE)	Total Time (Backsubst.)
16991101	310	9.68 sec.	0.13 sec.
17031491	132	3.31 sec.	0.13 sec.
17150031	126	3.43 sec.	0.11 sec.
16289	196	5.51 sec.	0.13 sec.
78937	190	5.51 sec.	0.15 sec.
16920245	142	3.31 sec.	0.12 sec.

TABLE 2

From the standpoint of execution time, the advantages of the backsubstitution scheme are quite apparent in the preceding table. It should be pointed out, however, that one must exercise some care so as not to be plagued by cumulative errors in the backsubstitution approach.

Another set of experiments of the same type was constructed in which each of the blocks was a randomly generated positive definite diagonal matrix. The constant column had entries randomly generated from a uniform distribution on the interval [-1,0] . The results of this set of runs are displayed in Table 3. They reflect the same advantage and the greater simplicity of the problems.

Initializer	(NULEMKE) Pivots	(NULEMKE) Time	(BACKSUBST.) Total Time
16471673	64	1.16 sec.	0.15 sec.
16507725	56	0.81 sec.	0.13 sec.
16539471	59	0.89 sec.	0.12 sec.
16568767	63	1.06 sec.	0.14 sec.
16597185	66	1.34 sec.	0.15 sec.
16630107	66	1.26 sec.	0.15 sec.
16668091	58	0.93 sec.	0.13 sec.
16716159	59	0.97 sec.	0.13 sec.

TABLE 3

5. A Sparse Problem of Medium Size

In [25], Hansen and Koopmans demonstrate that Scarf's fixed-point algorithm can be used to obtain an approximation to an optimal invariant capital stock. The paper also gives their computational experience with a particular numerical example. Later, Dantzig and Manne [16] studied the invariant capital stock problem from the standpoint of linear complementarity. Their approach entails the approximation of Hansen and Koopman's general concave utility function by one that is separable and piecewise linear. The formulation boils down to a linear complementarity problem of the form

$$q = \begin{pmatrix} -c \\ b \end{pmatrix} , \qquad M = \begin{pmatrix} 0 & c^T + D^T \\ -C & 0 \end{pmatrix} \qquad (5.1)$$

in which $D \geq 0$. Dantzig and Manne invoke the assumption that the linear programs

$$\left.\begin{array}{rl} \text{maximize} & c^T x \\ \text{subject to} & Cx \leq b \\ & x \geq 0 \end{array}\right\} \qquad (5.2)$$

$$\left.\begin{array}{rl} \text{minimize} & b^T y \\ \text{subject to} & (C + D)^T y \geq c \\ & y \geq 0 \end{array}\right\} \qquad (5.3)$$

each have finite optimal solutions. Doing so, they are able to prove that Lemke's algorithm applied to a problem (q,M) of the special structure shown in (5.1) will always produce a solution.

The numerical results reported by Dantzig and Manne were obtained from a special code implementing Lemke's algorithm written and executed by C. Engles. The actual complementarity problem solved was relatively small: M was of order 32. The adjusted running time for this code was about 1/8 of that reported by Hansen and Koopmans who used the fixed point approach. This outcome led Dantzig and Manne to suggest that "further work seems warranted in comparing the fixed-point and the complementarity algorithms on this class of models." (See [16, p. 16].)

Using the NULEMKE code and a finer grid, C. Engles treated the model again. This time, M was of order 320,

and only about 2% dense. The experiment had two purposes.
One was simply to check the code's performance on a linear
complementarity problem of medium size and low density. The
other was to look more closely at the relative merits of the
two solution strategies, in particular to test the Dantzig-
Manne formulation with a grid whose fineness and extent was
equivalent to the one used by Hansen and Koopmans. This was
accomplished by taking the piecewise linear approximation of
the separable additive objective function to consist of 100
pieces for each of the three summands. This approximation
contributed 300 of the 320 decision variables in the linear
complementarity problem. In each instance the solutions obtained were the same as those found by Hansen and Koopmans.
The pertinent computational facts are listed in Table 4.

α	Iterations	Execution Time
0.7	744	7.93 seconds
0.8	754	8.20 seconds
0.9	740	7.94 seconds

TABLE 4

The running times for these problems appear to be 4
times larger than the adjusted running times reported by
Hansen and Koopmans (14 minutes for all three problems combined on an IBM 1130). The author hesitates to draw too
strong an inference from this isolated set of nearly incom-
parable experiments; it may be said though that there is room
for doubt about the superiority of one method over the other
in this application.

6. Some Iterative Methods

The Gauss-Seidel method (see Varga [40, p. 58]) is a generic example of an _iterative_ method as the term is being used here. Even though it involves iterations, an algorithm such as Lemke's would be called a _direct_ method in this context.

The Gauss-Seidel method is used for solving the system of equations Ax = b where A is symmetric and positive definite, or equivalently for minimizing the quadratic function $Q(x) = \frac{1}{2} \langle x, Ax \rangle - \langle b, x \rangle$. The method works as follows: One chooses a starting trial solution $x^{(0)}$, k = 0 , and a stopping rule such as $\| r \|_\infty = \max_i |r_i| \leq \varepsilon$ where

$r = b - Ax$ is the _residual_ _vector_. If the stopping condition is met, then the current solution is accepted as the answer. Otherwise, one obtains $x^{(k+1)}$ from $x^{(k)}$ by successively solving the equations

$$x_i^{(k+1)} = - \frac{1}{a_{ii}} \left[\sum_{j<i} a_{ij} x_j^{(k+1)} + \sum_{j>i} a_{ij} x_j^{(k)} - b_i \right]$$
$$i = 1, \ldots, n \quad . \tag{6.1}$$

Under the stated hypothesis of symmetry and positive definiteness, the sequence of trial solutions $x^{(k)}$ converges to the unique solution x^∞ .

The Gauss-Seidel method has been used by several authors as the basis for an iterative algorithm to solve the analogous problem of minimizing a strictly convex quadratic

function on R_+^n ; this is equivalent to solving a linear
complementarity problem (q,M) having symmetric positive
definite M . Among those who have advanced this approach
are C. Hildreth [26], D. A. D'Esopo [19], and V. M. Fridman
and V. S. Chernina [22]. The Modified Gauss-Seidel Method
(as we shall call it) combines the cycle of one-dimensional
optimizations with a device that keeps the trial solutions
within the feasible domain, R_+^n . Starting from $x^{(0)} \geq 0$
and k = 0 one takes

$$x_i^{(k+\frac{1}{2})} = - \frac{1}{m_{ii}} \left[\sum_{j<i} m_{ij} x_j^{(k+1)} + \sum_{j>i} m_{ij} x_j^{(k)} + q_i \right] \qquad (6.2)$$

$$x_i^{(k+1)} = \max \{0, x_i^{(k+1/2)}\} \qquad (6.3)$$

for i = 1,...,n .

One of the notable features of this and some other iterative methods is that no transformation of the data takes
place. Furthermore, they can take advantage of initial estimates of the solution and of any inherent sparsity which, of
course, would be preserved throughout the procedure. (A direct method can rapidly create density out of sparsity.)

Successive overrelaxation (S.O.R.) and underrelaxation
(S.U.R.) are extensions of the Gauss-Seidel method. A relaxation parameter $\omega \in (0,2)$ is chosen. To solve Ax = b
with A symmetric and positive definite, one defines

$$x_i^{(k+\frac{1}{2})} = \frac{-1}{a_{ii}} \left[\sum_{j<i} a_{ij} x_j^{(k+1)} + \sum_{j>i} a_{ij} x_j^{(k)} - b_i \right] \qquad (6.4)$$

for each   i = 1,...,n   .   Then

$$x_i^{(k+1)} = x_i^{(k)} + \omega \left(x_i^{(k+\frac{1}{2})} - x_i^{(k)} \right) \quad . \tag{6.5}$$

When $\omega < 1$ ($\omega > 1$) one the method is called successive underrelaxation (overrelaxation). When $\omega = 1$, one re-covers the Gauss-Seidel method. The sequence of $x^{(k)}$ generated by this scheme converges to the solution. The rate of convergence depends on the choice of the relaxation parameter ω . Indeed, there is a formula for the optimal value of ω, but to apply this formula, one needs information that may be too costly to obtain to make its use worthwhile.

In the last few years a modification of S.O.R. for the linear complementarity problem (q,M) with symmetric posi-tive definite M has been described and numerically analyzed by C. W. Cryer [13], [14]. It appears that the basic idea for the algorithm can be traced to a 1941 paper of D. G. Christopherson [6] who used it in the context of lubrication problems.

The procedure combines the features of the Gauss-Seidel method with an overrelaxation step and a device for keeping all the iterates $x^{(k)}$ nonnegative. Starting from $x^{(0)} \geq 0$, k = 0   and $\omega \in (0,2)$, one uses the recursions

$$x_i^{(k+\frac{1}{2})} = - \frac{1}{m_{ii}} \left[\sum_{j<i} m_{ij} x_j^{(k+1)} + \sum_{j>i} m_{ij} x_j^{(k)} + q_i \right] \tag{6.6}$$

and

$$x_i^{(k+1)} = \max \left\{ 0, \ x_i^{(k)} + \omega \left(x_i^{(k+\frac{1}{2})} - x_i^{(k)} \right) \right\} \qquad (6.7)$$

Cryer's termination criterion is

$$\| r^{(k+1)} \|_\infty = \max_i | r_i^{(k+1)} | \leq \varepsilon$$

where

$$r_i^{(k+1)} = q_i + \sum_{j=1}^{i-1} m_{ij} x_j^{(k+1)} + \sum_{j=1}^{n} m_{ij} x_j^{(k)} \ .$$

He shows that the method is convergent.

7.   Computational Experience with Modified Point SOR

The Modified Point S.O.R. Method with Sacher's convergence criterion [36] (see Section 9 below) was coded by A.
Djang, who performed several sets of experiments using it.
The first set involved a totally dense randomly generated
symmetric positive definite matrix of order 50.   The experiment was designed to test the code and to show variation of
the optimal relaxation parameter with the tolerance ε .
The results are given in Table 5.

Tolerance	ω	Iterations	Execution Time
10^{-7}	0.80	454	1.514 sec.
10^{-7}	0.81	452	1.497 sec.
10^{-7}	0.82	452	1.497 sec.
10^{-7}	0.83	452	1.497 sec.
10^{-7}	0.84	454	1.497 sec.
10^{-7}	0.85	465	1.547 sec.
10^{-7}	0.86	466	1.547 sec.

TABLE 5

As the second experiment, the problem (q^*, M^*) of Table 1 was run. A tolerance of $\varepsilon = 10^{-7}$ was used. The relaxation parameter was varied from .8 to 1.8. The best result was as follows:

Tolerance ε	Optimal ω	Iterations	Execution Time
10^{-7}	0.96	335	6.722 sec.

TABLE 6

This result tends to support the contention that a good direct method like NULEMKE will outperform Modified Point S.O.R. on a medium size dense problem. Compare Table 1.

In [13, p. 443], C. Cryer reports on two sets of computational experiments with the Modified Point S.O.R. Method described in the previous section. Recall that the convergence criterion he uses is

$$\| r \|_{\infty} = \max_i |r_i| \leq \epsilon \quad . \tag{7.1}$$

The data for this set of experiments stem from the use of finite differences to solve the free boundary problem for infinite journal bearings. (See the formulation in [13].) In this case, the matrix M for the linear complementarity problem is of a very special kind. In particular, M is a tridiagonal Stieltjes matrix (see [40, p. 85]):

(i) $m_{ij} = 0$ if $|i - j| > 1$ } tridiagonal

(ii) $m_{ij} \leq 0$ if $i \neq j$
(iii) M is symmetric and positive definite } Stieltjes

when a matrix M has property (ii) and positive principal minors, it is said to be a Minkowski matrix. Thus a Stieltjes matrix is a symmetric Minkowski matrix.

The first set of experiments on which Cryer reports is arrived at exhibiting the influence of the relaxation parameter ω for a fixed problem of order $n = 63$. A. Djang has performed a similar experiment using the same data and the stopping rule $|x^T r| \leq \epsilon = 10^{-6}$. Djang's results are as follows:

Relaxation Parameter	Iterations	Execution Time
1.6	190	.74 sec.
1.7	130	.65 sec.
1.8	66	.62 sec.
1.9	73	.52 sec.
1.99	471	1.247 sec.

TABLE 7

These figures appear to be in agreement with those published by Cryer.

Next, a tridiagonal Minkowski problem of order 1023 was solved with Sacher's criterion and $\varepsilon = 10^{-7}$. A selection of the results are summarized in Table 8.

Relaxation Parameter	Iterations	Execution Time
1.986	645	7.970 sec.
1.990	721	8.852 sec.

TABLE 8

Further investigation of the relaxation parameters in the range $1.9862 \leq \omega \leq 1.9868$ yielded a run of 629 iterations and a running time of 7.854 sec. for $\omega = 1.9868$. In this case, one sees a fair amount of sensitivity to the relaxation parameter ω .

8.   A Direct Method for Tridiagonal Minkowski Matrices

Linear complementarity problems (q,M) in which M is a tridiagonal Minkowski matrix can be solved much more rapidly by a particular direct method especially designed to take advantage of their structure. The performance of this method is even better when the components of the vector q have a consecutive sign pattern, e.g.,

$$q_i < 0 \qquad i = 1,\ldots,m$$
$$q_i \geq 0 \qquad i = m+1,\ldots,n$$

(8.1)

The method which will be described very briefly here is based on an observation of Chandrasekaran [5] and is developed in full detail elsewhere [35], [12], [9].

When M is a Minkowski matrix and $q_i < 0$, the variable x_i must be positive in any (feasible) solution of the linear complementarity problem (q,M) . Indeed, we require

$$y_i = q_i + \sum_{j=1}^{n} m_{ij}x_j \geq 0 \qquad x_j \geq 0 \ (j = 1,\ldots,n)$$

Because $m_{ij} \leq 0$ $(i \neq j)$ and $m_{ii} > 0$, the only way to offset the negativity of q_i is to make $x_i > 0$, in which case $y_i = 0$ must hold in a solution of (q,M) . Thus, $q_i < 0$ implies x_i must be basic in any feasible solution. This can be accomplished by a principal pivot. An additional important fact is that once x_i becomes basic, it will never become nonbasic again. This in turn means that the ith column can be thereafter ignored, thereby assuring both computational and storage savings. The pivot operations have a very limited effect on the data, so as a result, it is possible to retain the original very low density and achieve a very efficient algorithm requiring very few arithmetic operations.

When the problem (q,M) is of this tridiagonal Minkowski type and (8.1) holds (which is the case for the infinite journal bearing problem), one must have

$$\sum_{j=1}^{m} m_{ij}x_j = -q_i \qquad i = 1,\ldots,m \quad . \qquad (8.2)$$

The system (8.2) is readily solved by matrix factorization. Its solution will be a positive vector. If substitution of $\bar{x}_1,\ldots,\bar{x}_m$ into the full system $y_i = q_i + \sum_{j=1}^{n} m_{ij}x_j$ ($i = 1,\ldots,n$) leads to $y_{m+1} < 0$, then a larger system of equations can be solved, and the earlier factorization can be exploited. (See [35], [36].)

A FORTRAN IV code embodying these ideas was written by R. S. Sacher. It was applied to a set of "journal bearing problems" of the type described above with the following results:

n	Execution Time in Milliseconds
63	< 1
127	16
255	49
511	216
1,023	1,015
2,047	3,560
4,095	13,561

TABLE 9

These running times clearly suggest the superiority of this direct algorithm over the Modified Point S.O.R. Method for problems of tridiagonal Minkowski type.

9. Block Tridiagonal MINKOWSKI Problems

The type of problem considered in the preceding section is actually of little consequence insofar as the journal bearing application is concerned because an analytic solution is available; hence a numerical solution is not so valuable. But such is not the case for the more realistic finite journal bearing problem. This free-boundary problem leads to a linear complementarity problem in which M is block tridiagonal and Minkowski (actually Stieltjes). Indeed, M is composed of blocks M_{ij} such that

  (i) M_{ii} is tridiagonal all i

 (ii) $M_{ij} \leq 0$ (elementwise), $i \neq j$

(iii)   M is Stieltjes

(iv) $M_{i,i+1}$ is diagonal

(v) $M_{ij} = 0$ if $|i - j| > 1$.

It follows that each main diagonal block M_{ii} is itself a Stieltjes matrix. See [40], [35], [11].

For this type of matrix structure, the observation made earlier that $x_h > 0$ whenever $q_h < 0$ is still valid, but unfortunately, the sparsity of M will not be preserved by making the corresponding principal pivot. The algorithm to be described here solves the problem in a manner which is analogous to Modified Point S.O.R. applied cyclically on the blocks. The method, called Modified Block S.O.R., is a hybrid consisting of iterative steps in which subproblems are solved directly.

Suppose M has m diagonal blocks M_{ii} . Let q and x be decomposed in conformity with the decomposition of M into blocks. Let $x^{(0)}$ be an initial nonnegative vector, let k = 0 and i = 1 . Choose a relaxation parameter $\omega \in (0,2)$ and a tolerance $\varepsilon > 0$ and then execute the following:

Step 1. Use the direct method of Section 7 to solve the linear complementarity problem

$$(q_i + \sum_{j<i} M_{ij} x_j^{(k+1)} + \sum_{j>i} M_{ij} x_j^{(k)}, M_{ii}) .$$

Let the solution be $x_i^{(k+\frac{1}{2})}$.

<u>Step 2.</u> Define

$$\omega_i^{(k+1)} = \max \{\bar{\omega}:\bar{\omega} \leq \omega, \ x_i^{(k)} + \bar{\omega}(x_i^{(k+\frac{1}{2})} - x_i^{(k)}) \geq 0\}$$

$$x_i^{(k+1)} = x_i^{(k)} + \omega_i^{(k+1)}(x_i^{(k+\frac{1}{2})} - x_i^{(k)}) \ .$$

<u>Step 3.</u> If i = m go to step 4. Otherwise return to Step 1
with i replaced by i + 1 .

<u>Step 4.</u> Define

$$S = \{(i,j) \mid (x_i^{(k+1)})_j > 0\} \cup \{(i,j) \mid (x_i^{(k+1)})_j = 0, \ (y_i^{(k+1)})_j < 0\} \ .$$

If $\max_{(i,j) \in S} |(y_i^{(k+1)})_j| \leq \varepsilon$, stop. An approximate solution
is at hand. If not, return to Step 1 with k replaced by
k+1 and i = 1 .

   It should be noticed that each of the subproblems to be
solved in Step 1 is of the tridiagonal Minkowski type considered in Section 7. The method is convergent under the stated
hypotheses.

   Table 10 below summarizes some computational experience
acquired in solving the finite journal bearing problem. In
each case, the starting trial solution was $x^{(0)} = 0$, and
the tolerance was $\varepsilon = 10^{-7}$. Running times are given for
relaxation parameters that were "experimentally optimal" to
within .02. The abbreviations MBSOR/F and MPSOR used in the

table stand for Modified Block S.O.R. with Factorization and
Modified Point S.O.R., respectively.

m	n=m^2	MBSOR/F		MPSOR		Time Ratio
		Iterations	Seconds	Iterations	Seconds	MPSOR:MBSOR/F
15	225	18	.133	45	.282	2.120
31	961	37	.881	87	2.296	2.606
63	3969	78	7.388	179	20.616	2.790

TABLE 10

It is interesting to contrast this performance with
some computational experience obtained in the process of numerically solving an entirely different underlying problem.
It was also of the free-boundary type, but arising in the
context of flow through a porous medium. The formulation is
given by Baiocchi et al. [2, p. 4]. Structurally the linear
complementarity problem that results from the finite difference approximation used is of the block tridiagonal Stieltjes
type to which the method of this Section applies.

In the single problem reported on in Table 11, the matrix M had 59 diagonal blocks each of order 39, so its order n was 2301. The starting point and stopping criterion
were as above: $x^{(0)} = 0$ and $\varepsilon = 10^{-7}$.

	Iterations	Time	ω_{exp}
MBSOR/F	106	7.953	1.78
MPSOR	159	9.684	1.86

TABLE 11

In running this problem, it was observed that the two methods performed with about equal efficiency whereas MBSOR/F had a more decided edge in the finite journal bearing problem. One respect in which the linear complementarity problems differed was in the percentage of x-variables that became positive in the final accepted solution. In the journal bearing problem a little more than 50% of the x_i were positive while in the porous flow problem this percentage was very much higher. It was conjectured that the more a problem behaves like a system of linear equations and less like a linear complementarity problem, the less superior the MBSOR/F method would perform by comparison with MPSOR. The results of the following experiment tend to support this conjecture.

Define the 900 X 900 block tridiagonal Stieltjes matrix $M = (M_{ij})$ in which

$$M_{ii} = T \qquad i = 1,\ldots,30$$

$$M_{i,j} = -I \qquad |i - j| = 1$$

$$M_{i,j} = 0 \qquad |i - j| > 1$$

and

$$
T = \begin{bmatrix}
4 & -1 \\
-1 & 4 & -1 \\
 & -1 & 4 & -1 \\
 & & & \ddots & \ddots & \ddots \\
 & & & & \ddots & \ddots & \ddots \\
 & & & & & \ddots & \ddots & \ddots \\
 & & & & & & \ddots & \ddots & \ddots \\
 & & & & & & & \ddots & \ddots & \ddots \\
 & & & & & & & & \ddots & \ddots & \ddots \\
 & & & & & & & & -1 & 4 & -1 \\
 & & & & & & & & & -1 & 4
\end{bmatrix}
$$

The matrices T and I are of order 30. Next let q^t be a column vector in R^{900} having its first 30t components equal to -3 and its remaining $30(30 - t)$ components equal to $+1$. As t increases from 1 to 30 , (q^t, M) becomes more like a linear system in the sense that more x-variables must be positive as more components of the constant vector q become negative. The outcome of this experiment appears in Table 12.

t	Positive Var.in Sol.	MBSOR/F Iter.	Sec.	ω_{exp}	MPSOR Iter.	Sec.	ω_{exp}	Time Ratio MPSOR:MBSOR/F
1	60	7	.099	1.08	19	.449	1.20	4.535
2	118	14	.216	1.26	32	.732	1.40	3.389
3	174	20	.349	1.40	42	.998	1.50	2.860
6	336	36	.765	1.58	60	1.431	1.68	1.871
9	480	50	1.148	1.66	79	1.880	1.76	1.638
12	610	60	1.580	1.72	89	2.113	1.78	1.337
30	900	97	2.995	1.74	124	2.916	1.82	.989

TABLE 12

From this experiment, we make the tentative inference that the Modified Block S.O.R. algorithm is only at an advantage when the complementarity aspect of the problem is nontrivial. Our experience is in line with the known performance characteristics of Block versus Point S.O.R. methods. Furthermore, it has been suggested by G. H. Golub that a suitable normalization can bring about even greater efficiency in the solution of the subproblems of the block method. See Varga [40, p. 199].

We are presently engaged in attempting to broaden the scope of the Modified Block S.O.R. approach. For example, we hope to find a way to solve the type of problem considered and treated so successfully by Cea and Glowinski [4, p. 26].

REFERENCES

[1] Baiocchi, C., "Sur Quelques Problemes à Frontiere
     Libre," Asterisque 2 et 3, 1973 (Societe Mathematique
     de France), pp. 70-85.

[2] Baiocchi, C., Cominciolo, V., Guerri, L., and Volpi, G.,
     "Free Boundary Problems in the Theory of Fluid Flow
     Through Porous Media: A Numerical Approach," Calcolo
     10 (1973), 1-86.

[3] Cea, J., "Recherche Numérique d'un Optimum Dans un
     Espace Produit," in N. N. Moiseev, ed., Colloquium on
     Methods of Optimization, Lecture Notes in Mathematics
     No. 112, Springer-Verlag, Berlin 1970, pp. 33-50.

[4] Cea, J., et Glowinski, R., "Sur des Methods d'Optimisa-
     tion par Relaxation," Revue, Francaise d'Automatique,
     Informatique et Recherche Operationnelle R-3 (1973) 5-32.

[5] Chandrasekaran, R., "A Special Case of the Complemen-
     tary Pivot Problem," Opsearch 7 (1970), 263-268.

[6] Christopherson, D. G., "A New Mathematical Method for
     the Solution of Film Lubrication Problems," Proceedings,
     Institute of Mechanical Engineers 146 (1941), 126-135.

[7] Cottle, R. W., "On a Problem in Linear Inequalities,"
     J. London Math. Soc. 43 (1968), 378-384.

[8] Cottle, R. W., "Solution Rays for a Class of Complemen-
     tary Problems," Mathematical Programming Study 1,
     Pivoting and Extensions (M. L. Balinski, ed.), Amster-
     dam, November 1974, 59-70.

[9] Cottle, R. W., "Complementarity and Variational Prob-
     lems," to appear in Symposia Mathematica.

[10] Cottle, R. W., and Dantzig, G. B., "Complementary Pivot
     Theory of Mathematical Programming," Linear Algebra and
     Its Application 1 (1968), 103-105.

[11] Cottle, R. W., Golub, G. H., and Sacher, R. S., "On the
      Solution of Large, Structured Linear Complementarity
      Problems: III," Technical Report 73-8, Department of
      Operations Research, Stanford University, June 1974.

[12] Cottle, R. W., and Sacher, R. S., "On the Solution of
      Large, Structured Linear Complementarity Problems: I,"
      Technical Report 73-4, Department of Operations Re-
      search, Stanford University, April 1973.

[13] Cryer, C. W., "The Method of Christopherson for Solving
      Free Boundary Problems for Infinite Journal Bearings by
      Means of Finite Differences," Mathematics of Computa-
      tion 25 (1971), 435-443.

[14] Cryer, C. W., "The Solution of a Quadratic Programming
      Problem Using Systematic Overrelaxation," SIAM J.
      Control 9 (1971), 385-392.

[15] Dantzig, G. B., Linear Programming and Extensions,
      Princeton University Press, Princeton, N.J., 1963.

[16] Dantzig, G. B., and Manne, A. S., "A Complementarity
      Algorithm for an Optimal Capital Path with Invariant
      Proportions," Technical Report 74-1, January 1974
      (Revised March 1974), Department of Operations Research,
      Stanford University.

[17] DeDonato, O., and Franchi, A., "A Modified Gradient
      Method for Finite Element Elastoplastic Analysis by
      Quadratic Programming," Technical Report, Instituto di
      Scienze e Technica delle Costruzioni, Politecnico di
      Milano, July 1972.

[18] DeDonato, O., and Maier, G., "Mathematical Programming
      Methods for the Inelastic Analysis of Reinforced Con-
      crete Frames Allowing for Limited Rotation Capacity,"
      International Journal for Numerical Methods in Engi-
      neering 4 (1972), 307-329.

[19] D'Esopo, D. A., "A Convex Programming Procedure," Naval
      Research Logistics Quarterly 6 (1959), 33-42.

[20] Eaves, B. C., "The Linear Complementarity Problem,"
      Management Science 17 (1971), 612-734.

[21] Eaves, B. C., "On the Basic Theorem of Complementarity," Mathematical Programming 1 (1971), 68–75.

[22] Fridman, V. M., and Chernina, V. S., "An Iteration Process for the Solution of the Finite-Dimensional Contact Problem," U.S.S.R. Comp. Math. and Math. Phys. 8 (1967), 210–214.

[23] Garcia, C. B., "Some Classes of Matrices in Linear Complementarity Theory," Mathematical Programming 5 (1973), 299–310.

[24] Glowinski, R., "Sur la Minimization, par Surrelaxation Avec Projection de Fonctionnelles Quadratiques dans les Espaces de Hilbert," C. R. Acad. Sci., Paris 276 (1973), 1421–1423.

[25] Hansen, T., and Koopmans, T. C., "On the Definition of a Capital Stock Invariant under Optimization," J. Economic Theory 5 (1972), 487–523.

[26] Hildreth, C., "A Quadratic Programming Procedure," Naval Research Logistics Quarterly 4 (1957), 79–85.

[27] Ingleton, A. W., "A Problem in Linear Inequalities," Proc. London Math. Soc. 16 (1966), 519–536.

[28] Karamardian, S., "Duality in Mathematical Programming," Doctoral dissertation, University of California, Berkeley, 1966.

[29] Kluge, R., "Folgen und Interationsverfahren bei Folgen Nichtlinearer Variationsungleichungen," in M. Kucera, ed., Theory of Nonlinear Operators, Academic Press, New York, 1973, pp. 39–47.

[30] Lemke, C. E., "Bimatrix Equilibrium Points and Mathematical Programming," Management Science 11 (1965), 681–689.

[31] Macchi, G., "Limit States Design of Statistically Indeterminate Structures Composed of Linear Members," Studi e Rendiconti del Corso di Perfezionamento di Costruzioni in Cemento Armato, Politecnico di Milano, 6 (1969), 151–191.

[32]   Murty, K. M., "On the Number of Solutions of the Com-
       plementarity Problem and Spanning Properties of Comple-
       mentary Cones," Linear Algebra and Its Appl. 5 (1972),
       65-108.

[33]   Niccolucci, F., "On the Equivalence Among Integer Pro-
       grams, Nonconvex Real Programs, Special Complementarity
       Problems, Variational Inequalities and Fixed-Point Pro-
       blems," Pubblicazioni del Dipartimento di Ricerca Oper-
       ativa e Scienze Statistiche, Serie A, No. 8, Universita
       di Pisa, July 1973.

[34]   Polizzotto, C., "Some Minimum Theorems for Rates His-
       tories in Structural Elastoplasticity," Technical Re-
       port SISTAR-74-OMS-4, Jan. 1974, Pubblicazione dell'
       Instituto di Scienza e Technica delle Costruzioni,
       Facolta di Architettura, Universita di Palermo, No. 101.

[35]   Sacher, R. S., "On the Solution of Large, Structured
       Linear Complementarity Problems: II," Technical Report
       73-5, Department of Operations Research, Stanford
       University, September 1973.

[36]   Sacher, R. S., "On the Solution of Large, Structured
       Linear Complementarity Problems," Doctoral dissertation,
       Department of Operations Research, Stanford University,
       1974.

[37]   Saigal, R., and Simon, C., "Generic Properties of the
       Complementarity Problem," Mathematical Programming 4
       (1973), 324-335.

[38]   Samelson, H., Thrall, R. M., and Wesler, O., "A Parti-
       tion Theorem for Euclidean N-Space," Proc. Amer. Math.
       Soc. 9 (1958), 805-807.

[39]   Scarf, H., "The Approximation of Fixed Points of a Con-
       tinuous Mapping," SIAM J. Appl. Math. 15 (1967), 1328-
       1343.

[40]   Varga, R. S., Matrix Iterative Analysis, Prentice Hall,
       Englewood Cliffs, New Jersey, 1962.

[41]   Watson, L. T., "A Variational Approach to the Linear
       Complementarity Problem," Doctoral dissertation, Univ-
       ersity of Michigan, Ann Arbor, 1974.

Union Jack Triangulations

Michael J. Todd

ABSTRACT

    Triangulations of the simplex and the cube given by
Whitney and Tucker are reexamined with rexpect to their possible use for fixed point computation. The traingulations
are extended by the use of an auxiliary dimension to allow
continuous refining of the grid size as in Eaves' K_1 and
K_2 and Eaves and Saigal's K_3 .

1. Introduction

The simplest triangulation of a simplex is the well-known barycentric subdivision, and used for example by Shapley [10]. Unfortunately, iterating this operation to get arbitrarily fine traingulations leads to simplices of increasingly long, skinny shapes. This "flatness" [4] or lack of "fullness" [11] was recognized to be a problem in iteration theory well before the possibility of computing fixed points arose. The question of finding a simple construction of arbitrarily fine triangulations of the simplex with bounded flatness seems to have been posed by Brouwer, and was first answered by Freudenthal [4]. Freudenthal's construction is the basis of the standard triangulation of the cube or simplex (see problem 3, page 140 of [9], attributed to Tucker; Kuhn [6] and [7]; Hansen [5] and Eaves [1]). This triangulation is also the basis of Eaves' K_1 and K_2 [2] and Eaves and Saigal's K_3 [3]. We will describe here an alternative triangulation of the cube and simplex. The triangulation of the cube is the centrally symmetric subdivision due to Tucker (problem 4, page 140 of [9]) or barycentric subdivision (Whitney [11] page 358). The related triangulation of the simplex is described by Whitney ([11], pages 358-60). We will also extend these triangulations with an auxiliary dimension to obtain the advantages of a continuously refined grid size. The latter have a slightly simpler characterization and pivot rules than Eaves' K_2

and Eaves' and Saigal's K_3 . For 2 dimensions, Kuhn [8]
has developed a continuously refining triangulation which
seems to be equivalent to ours. The two-dimensional case
leads to the suggestive naming of the triangulations in the
title (see Figure 1).

2. Definitions

     An (open) j-dimensional simplex σ is the relative in-
terior of the convex hull of j+1 affinely independent
points $y^o, y^1, \ldots, y^j$, called its vertices. We write
$\sigma = \langle y^o, \ldots, y^j \rangle$. A face of σ is a simplex determined by
any subset of its vertices - thus σ is a face of itself.
All simplices have a given ordering of their vertices. If
$\sigma = \langle y^o, y^1, \ldots, y^j \rangle$, the edge vectors of σ are
$v^i = y^i - y^{i-1}$, $1 \le i \le j$. The diameter of σ is
$\sup \{ \| x-z \| \, | x,z \in \sigma \} = \max \{ \| y^i - y^k \| \, | 0 \le i \le k \le j \}$.

     If T is a collection of n-simplices, T^+ denotes the
collection of all faces of members of T . The mesh of T
is $\sup \{\text{diameter of } \sigma | \sigma \in T\}$. T is a triangulation of
E if T^+ is a partition of E .

     Let A denote $\{-1,+1\}$. All points of A^{n+1} and
R^{n+1} will have coordinates indexed 0 through n . Points
of R^{n+2} will have coordinates indexed -1 through n .
If a matrix has n+1 (n+2) rows or columns, they will be indexed 0 through n (-1 through n). Σ_n and Σ_{n+1} denote
the groups of permutations on $\{1,2,\ldots,n\}$ and
$\{0,1,2,\ldots,n\}$ respectively.

3.   The Triangulation of R^n

We will define a triangulation $J_1 = J_1(\delta)$ of R^n depending on a scale factor $\delta \in R$. Let $J_1^o = \{y \in R^n | y_i/\delta$ is integral for $1 \leq i \leq n\}$ be the set of vertices of J_1 , and $J_1^{oc} = \{y \in J_1^o | y_i/\delta$ is odd for $1 \leq i \leq n\}$ be the set of central vertices. Note that if $x \in R^n$, $|x_i - y_i| \leq \delta$ for $1 \leq i \leq n$ for some $y \in J_1^{oc}$.

A simplex $\sigma = \langle y^o, \ldots, y^n \rangle$ is a member of J_1 if there is a triple (y, π, s) in $J_1^{oc} \times \Sigma_n \times A^n$ such that

$$y^o = y$$

$$(3.1)$$

$$y^i = y^{i-1} + \delta s_{\pi(i)} e^{\pi(i)} , \qquad 1 \leq i \leq n$$

where e^j is the jth unit vector in R^n . We also write $\sigma = (y, \pi, s)$. The triangulation J_1 for $n = 2$, $\delta = 1$ is shown in Figure 1.

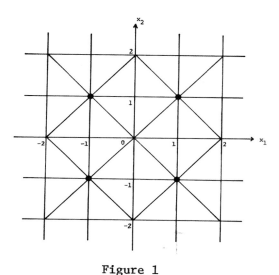

Figure 1

J_1^{0c} - heavy dots

Lemma 3.2. J_1 is a triangulation of R^n .

Proof. Let $x \in R^n$ be arbitrary, and let y be a closest central vertex to x . By examining $|x_i - y_i|$, $1 \leq i \leq n$, a permutation $\pi \in \Sigma_n$ can be found so that

$$\delta \geq |x_{\pi(1)} - y_{\pi(1)}| \geq \cdots \geq |x_{\pi(n)} - y_{\pi(n)}| \geq 0 . \quad (3.2)$$

For $1 \leq i \leq n$, pick $s_{\pi(i)} \in A$ so that $s_{\pi(i)}(x_{\pi(i)} - y_{\pi(i)}) \geq 0$. Then

$$\delta \geq s_{\pi(1)}(x_{\pi(1)} - y_{\pi(1)}) \geq \cdots \geq s_{\pi(n)}(x_{\pi(n)} - y_{\pi(n)}) \geq 0 \quad .$$

$$(3.3)$$

Denote the terms above $\alpha_0 = \delta$, $\alpha_1, \ldots, \alpha_n$, $\alpha_{n+1} = 0$, and let $\beta_i = (\alpha_i - \alpha_{i+1})/\delta$ for $0 \leq i \leq n$. It is readily seen that

$$x = \sum_{i=0}^{n} \beta_i y^i, \quad \beta_i \geq 0, \quad 0 \leq i \leq n, \quad \text{and} \quad \sum_{i=0}^{n} \beta_i = 1 \quad (3.4)$$

where $y^0, \ldots, y^n$ are given by (3.1). Thus $x \in \bar{\sigma}$ for $\sigma = (y, \pi, s) \in J_1$. Since $\bar{\sigma}$ is covered by σ and its faces, R^n is covered by J_1^+ . Moreover, since (3.4) implies (3.3), with positive β_i's corresponding to strict inequalities in (3.3), it follows that each $x \in R^n$ lies in a unique (open) simplex of J_1^+ .

For use in fixed point computation, the pivot rules of J_1 are essential. Let $\sigma = (y, \pi, s) = y^0, \ldots, y^n$, and let $\tau = (z, \rho, t) \in J_1$ contain all vertices of σ except y^i . Then τ is obtained from σ by the rules of Table 1.

Table 1

	z	ρ	t
$i = 0$	$y + 2\delta s_{\pi(1)}e^{\pi(1)}$	π	$s - 2s_{\pi(1)}e^{\pi(1)}$
$0 < i < n$	y	$(\pi(1),\ldots,\pi(i+1),\pi(i),\ldots,\pi(n))$	s
$i = n$	y	π	$s - 2s_{\pi(n)}e^{\pi(n)}$

4. The Triangulation of S^n

We will use J_1 to triangulate $S^n = \{x \in R^{n+1} | x \geq 0$ $\sum_{i=0}^{n} x_i = 1\}$. Throughout, the δ of section 3 is set to $1/m$, with m a positive integer. For m a power of 2, Whitney's standard subdivision [11] is recovered.

Let C^n denote $\{x \in R^n | 1 \geq x_1 \geq , \ldots , \geq x_n \geq 0\}$. A linear homeomorphism h from C^n to S^n is obtained as follows. Denote by U and Q the $(n+1) \times n$ and $n \times (n+1)$ matrices

$$
U = \begin{pmatrix} -1 & 0 & \cdots & 0 \\ 1 & -1 & & 0 \\ 0 & 1 & & 0 \\ \cdot & \cdot & & \cdot \\ \cdot & \cdot & & \cdot \\ \cdot & \cdot & & -1 \\ 0 & 0 & & 1 \end{pmatrix} , \quad Q = \begin{pmatrix} 0 & 1 & 1 & \cdots & 1 \\ 0 & 0 & 1 & \cdots & 1 \\ \cdot & \cdot & \cdot & & \cdot \\ \cdot & \cdot & \cdot & & \cdot \\ \cdot & \cdot & \cdot & & \cdot \\ 0 & 0 & 0 & \cdots & 1 \end{pmatrix} .
$$

Let f^j be the jth unit vector in R^{n+1} . Then $h(c) = f^0 + Uc$, $h^{-1}(s) = Qs$ provides the required homeomorphism. Any triangulation of C^n yields a triangulation of S^n by applying h to each simplex.

Lemma 4.1. The subset of J_1 consisting of all n-simplices meeting C^n triangulates C^n .

Proof. Let J_1' be the subset of J_1 meeting C^n , and $J_1'^+$ the collection of all faces of members of J_1' . It is clear that $J_1'^+$ covers C^n . Moreover, any $x \in C^n - \text{int } C^n$ satisfies (3.3) with at least one equality, and so cannot lie in any $\sigma \in J_1'$. It follows that each $\sigma \in J_1'$ lies wholly in C^n , and therefore $J_1'^+$ is a partition of C^n .

Lemma (4.1) and the homeomorphism h give a triangulation J_2 of S^n . Let $J_2^o = \{y \in S^n \mid my_i$ is integral for $0 \le i \le n\}$ and $J_2^{oc} = \{y \in J_2^o \mid m - my_i$ is odd for $i = 0$ and n , even for $0 < i < n\}$.

A simplex $\sigma = \langle y^o, \ldots, y^n \rangle$ is a member of J_2 if each $y^i \in J_2^o$ and there is a triple (y, π, s) in $J_2^{oc} \times \Sigma_n \times A^n$ such that:

$$y^o = y$$

$$y^i = y^{i-1} + (1/m) s_{\pi(i)} u^{\pi(i)}, \qquad 1 \le i \le n \ . \quad (4.2)$$

where u^j is the jth column of U . Pivot rules for J_2 are given by Table 1, except that in the top left corner z is $y + (2/m) s_{\pi(1)} u^{\pi(1)}$.

We now relate J_2 to Whitney's standard subdivision [11]. Given a simplex $\sigma = \langle y^o, \ldots, y^n \rangle$, let $J^o =$ $\{y^{ij} = \frac{1}{2} y^i + \frac{1}{2} y^j | 0 \leq i \leq j \leq n\}$, and define a partial order on J^o by $y^{ij} \leq y^{k\ell}$ iff $i \leq k$ and $j \geq \ell$. Let $\tau \in J$ if τ is a maximal chain in J^o . Clearly every $\tau \in J$ has y^{on} as first vertex, and its edge vectors lie in the set $\{ \pm \frac{1}{2} v^1, \ldots, \pm \frac{1}{2} v^n \}$, where $v^1, \ldots, v^n$ are the edge vectors of σ . Whitney's subdivision is obtained by iterating this operation on each simplex of J , each simplex thus obtained, and so on. If $\sigma = S^n$ and the subdivision is repeated p times, what is obtained is precisely J_2 , with $m = 2^p$. To see this, note that the set of first vertices of simplices of the pth subdivision is precisely J_2^{oc} ; this is easily proved by induction. Also, the edge vectors of σ are the columns of U , and so each simplex of Whitney's subdivision lies in J_2 . Since J_2 is a triangulation of S^n , the equivalence follows.

For $n = 2$ and $m = 2$ and 4 , J_2 is shown in Figure 2.

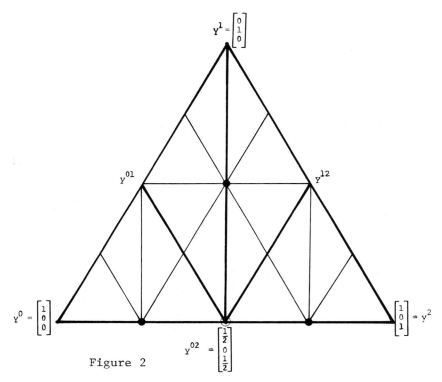

Figure 2

For m = 2, J_2 is the set of heavily outlined trianges,
and J_2^{0c} is the circled vertex. For m = 4, J_2 is the
set of all trianges and J_2^{0c} is the set of heavy dots.

Let us note some comparisons of J_1 and J_2 with

standard triangulations. First, the mesh of J_1 is $\delta\sqrt{n}$,

that of J_2 $\sqrt{4n-2}/m$. The former compares favorably with

the standard triangulation, while the latter is higher by a

factor of about 2. The pivot rules seem about equally sim-

ple - in J_1 and J_2 a sign vector must be stored, but π

is updated less often. An interesting feature of J_1 and

J_2 is that all vertices occur in the same position in each simplex in which they occur; for J_1 , this position corresponds to the number of even coordinates in $(1/\delta)y$.

5. Triangulations of $R^n \times (0,1]$ and $S^n \times (0,1]$.

We now extend J_1 and J_2 by adding an auxiliary dimension to obtain the advantages of continuously refined grid size as in Eaves' K_1 and K_2 [2] and Eaves' and Saigal's K_3 [3].

$R^n \times (0,1]$ is considered as a subset of R^{n+1} , with the 0th coordinate lying in the auxiliary dimension. A triangulation of $R^n \times (0,1]$ yields one of $R^n \times [0,\infty)$ by means of the homeomorphism $h(y) = ([(1-y_0)/y_0],y_1,\ldots,y_n)^T$ for $y \in R^n \times (0,1]$.

Denote by J_3^o the set of vertices $\{y \in R^{n+1}|y_0 = 2^{-k}$ for $k = 0,1,\ldots$ and y_i/y_0 intergral for $1 \leq i \leq n\}$. Such a vertex has depth k if $y_0 = 2^{-k}$. Let $J_3^{oc} = \{y \in J_3^o|y_i/y_0$ is odd for $1 \leq i \leq n\}$ be the sets of central vertices. For each central vertex of depth $k \geq 1$ we can associate the closest central vertex of depth $k-1$ as follows.

Define ν : $J_3^{oc} \to A^{n+1}$ by $\nu_i(y) = -1$ or $+1$ according as y_i/y_0 is 1 or 3 mod 4 .

Then $y^1 = y - y_0 \, \nu(y) \in J_3^{oc}$ has depth $k-1$ and is closest to y .

A simplex $\sigma = \langle y^{-1}, y^0, \ldots, y^n \rangle$ is a member of J_3 if there is a triple $(y, \pi, s) \in J_3^{oc} \times \Sigma_{n+1} \times A^{n+1}$ where y has depth at least one, such that:

$$y^{-1} = y$$

$$y^i = y^{i-1} + y_0 s_{\pi(i)} f^{\pi(i)} \qquad 0 \le i < j = \pi^{-1}(0)$$

(5.1)

$$y^j = y^{j-1} - y_0 \sum_{\ell=j}^{n} \nu_{\ell(\pi)} f^{\pi(\ell)}$$

$$y^k = y^{k-1} + 2y_0 \nu_{\pi(k)} f^{\pi(k)} \qquad j < k \le n \; ,$$

where $\nu = \nu(y)$. Recall that f^j is the jth unit vector in R^{n+1} , $0 \le j \le n$. Note that σ does not depend on $s_{\pi(j)}, \ldots, s_{\pi(n)}$, nor on $\nu_{\pi(0)}, \ldots, \nu_{\pi(j)}$. We also write $\sigma = (y, \pi, s)$.

Lemma 5.2. J_3 is a triangulation of $R^n \times (0,1]$.

Proof. Let x be any point of $R^n \times (0,1]$ with $2^{-k+1} \ge x_0 > 2^{-k}$. Let y be a closest central vertex of depth k to x . By examining $|x_i - y_i|$, $0 \le i \le n$, a permutation $\rho \in \Sigma_{n+1}$ can be found such that

$$y_0 \geq |x_{\rho(0)} - y_{\rho(0)}| \geq \ldots, \geq |x_{\rho(n)} - y_{\rho(n)}| \geq 0 .$$

Let $j = \rho^{-1}(0)$. $\pi \in \Sigma_{n+1}$ will agree with ρ on 0
through j , and is chosen together with $s \in A^{n+1}$, so
that

$$y_0 \geq s_{\pi(0)}(x_{\pi(0)} - y_{\pi(0)}) \geq \ldots \geq s_{\pi(j-1)}(x_{\pi(j-1)} - y_{\pi(j-1)})$$
$$\geq x_0 - y_0 \geq$$
$$v_{\pi(j+1)}(x_{\pi(j+1)} - y_{\pi(j+1)}) \geq \ldots \geq v_{\pi(n)}(x_{\pi(n)} - y_{\pi(n)}) \qquad (5.3)$$
$$\geq y_0 - x_0 ,$$

where $v = v(y)$.

Let $\alpha_{-1} = y_0, \alpha_0, \ldots, \alpha_j = x_0 - y_0, \alpha_{j+1}, \ldots, \alpha_n, \alpha_{n+1} = y_0 - x_0$ denote the terms above, and define
$\beta_i = (\alpha_i - \alpha_{i+1})/y_0$ for $-1 \leq i < j$ and $\beta_k = (\alpha_k - \alpha_{k+1})/2y_0$
for $j \leq k \leq n$. Then $\beta \geq 0$, and
$$\sum_{\ell=-1}^{n} \beta_\ell = 1 - (x_0 - y_0)/2y_0 - (y_0 - x_0)/2y_0 = 1 , \text{ and also}$$

$$x = \sum_{\ell=-1}^{n} \beta_\ell y^\ell \qquad (5.4)$$

where $y^{-1}, \ldots, y^n$ are given by (5.1). Thus $x \in \bar{\sigma}$ for
$\sigma = (y, \pi, s) \in J_3$. Since $\bar{\sigma}$ is covered by σ and its
faces, $R^n \times (0,1]$ is covered by J_3^+ . Moreover, since

(5.4) implies (5.3), with positive β_ℓ's corresponding to

strict inqualities in (5.3), it follows that each

$x \in R^n \times (0,1]$ lies in a unique simplex of J_3^+ .

The pivot rules of J_3 are shown in Table 2. Here the

$i = -1$ and $i = n$ cases take precedence over those for

$i = j$ or $i = j-1$ when they occur together. Also, although

s is "free" while ν is determined by y , it is conve-

nient to store together the relevant components of s and

ν in a vector $b \in R^n$. Let $b_{\pi(i)} = s_{\pi(i)}$ for $0 \le i < j$

and $b_{\pi(k)} = \nu_{\pi(k)}$ for $j < k \le n$. Let $\sigma' = (z,\ell,t) \in J_3$

contain all vertices of $\sigma = (y,\pi,s) \in J_3$ except y^i , and

let b be obtained from π, s and $\nu(y)$ as above, and c

from ρ, t and $\nu(z)$ similarly. Then z, ρ and c are

obtained from y, π and b as in Table 2, where $j = \pi^{-1}(0)$.

As in section 4, we can use J_3 and the homeomorphism

h to triangulate $S^n \times (0,1]$. We will consider the latter

a subset of R^{n+2} , where the -1st coordinate lies in the

auxiliary dimension. The following result is proved exactly

as lemma 4.1.

<u>Lemma 5.5.</u> The subset of J_3 consisting of all n-simplices

meeting $C^n \times (0,1]$ triangulates $C^n \times (0,1]$.

Using h and lemma 5.5, we obtain the triangulation

J_4 of $S^n \times (0,1]$. Let J_4^o {$y \in R^{n+2} | y_{-1} = 2^{-k}$ for

$k = 0,1,\ldots,$ $(y_0,\ldots,y_n) \in S^n$ and y_i/y_{-1} is integral for

$0 \le i \le n$} be the set of vertices of J_4 . Such a vertex

Table 2

		z	ρ	c
$i = -1$	$j = 0$	$y - y_0(-1,b)$	$(\pi(1)\ldots\pi(n)\pi(0))$	b
	$j > 0$	$y+2y_0 b_{\pi(0)} f^{\pi(0)}$	π	$b-2b_{\pi(0)}e^{\pi(0)}$
	$0 \leq i < j-1$	y	$(\pi(0)\ldots\pi(i+1)\pi(i)\ldots\pi(n))$	b
$i=j-1$	$b_{\pi(j-1)} = \nu_{\pi(j-1)}$	y	$(\pi(0)\ldots\pi(j)\pi(j-1)\ldots\pi(n))$	b
	$b_{\pi(j-1)} = -\nu_{\pi(j-1)}$	y	$(\pi(0)\ldots\pi(j-2),\pi(j)\ldots\pi(n)\pi(j-1))$	$b-2b_{\pi(j-1)}e^{\pi(j-1)}$
	$j \leq i < n$	y	$(\pi(0)\ldots\pi(i+1),\pi(i)\ldots\pi(n))$	b
$i = n$	$j = n$	$y + \frac{1}{2} y_0(-1,b)$	$(\pi(n),\pi(0),\ldots,\pi(n-1))$	b
	$j < n$	y	$(\pi(0)\ldots\pi(j-1),\pi(n),\pi(j)\ldots\pi(n-1))$	$b-2b_{\pi(n)}e^{\pi(n)}$

has depth k if $y_{-1} = 2^{-k}$. Let $J_4^{0c} = \{y \in J_4^0 | y_i/y_{-1}$ is

odd for $i = 0$ and n, even for $0 < i < n\}$ be the set of

central vertices. Define $v': J_4^0 \to A^{n+1}$ by $v_n'(y) = -1$

or $+1$ according as y_n/y_0 is 1 or 3 mod 4, and

$v_i'(y) = v_{i+1}'(y)$ or $-v_{i+1}'(y)$ according as y_i/y is 0 or

2 mod 4 for $1 \leq i < n$, and $v'(y) = -1$.

Let W be the $(n+2) \times (n+1)$ matrix with columns $w^0, w^1, \ldots, w^n$.

$$
\begin{pmatrix}
1 & 0 & 0 & \cdots & 0 \\
0 & -1 & 0 & \cdots & 0 \\
0 & 1 & -1 & \cdots & 0 \\
0 & 0 & 1 & \cdots & 0 \\
\cdot & \cdot & \cdot & & \cdot \\
\cdot & \cdot & \cdot & & \cdot \\
\cdot & \cdot & \cdot & & -1 \\
0 & 0 & 0 & \cdots & 1
\end{pmatrix}
$$

A simplex $\sigma = \langle y^{-1}, \ldots, y^n \rangle$ is a member of J_4 if

each $y^i \in J_4^0$ and there is a triple $(y, \pi, s) \in J_4^{0c} \times \Sigma_{n+1} \times A^{n+1}$ where y has depth at least one, such that:

$$y^{-1} = y$$

$$y^i = y^{i-1} + y_0 s_{\pi(i)} w^{\pi(i)} \qquad 0 \leq i < j = \pi^{-1}(0)$$

$$y^j = y^{j-1} - y_0 \sum_{\ell=j}^{n} v_{\pi(\ell)} w^{\pi(\ell)} \qquad\qquad (5.6)$$

$$y^k = y^{k-1} + 2y_0 v_{\pi(k)} w^{\pi(k)} \qquad j < k \leq n$$

where $\nu = \nu'(y)$. We also write $\sigma = (y,\pi,s)$.

Pivot rules for J_4 are also obtained from Table 2, except that w^ℓ replaces f^ℓ everywhere.

Figure 3 shows J_3 for $n = 1$. Figure 4 shows all simplices of J_3 for $n = 2$ of the form $\sigma = (y,\pi,s)$ with $y = (\frac{1}{2},\frac{1}{2},\frac{1}{2})$. Figure 5 shows all simplices of J_4 for $n = 2$ of the form $\sigma = (y,\pi,s)$ with $y = (\frac{1}{2},\frac{1}{2},0,\frac{1}{2})$.

The triangulations J_3 and J_4 satisfy conditions equivalent to (5.1)-(5.3) of [3] and (a)-(c) of section 6 of [2] and thus can be used instead of K_1,K_2, and K_3 in the fixed points algorithms of Eaves and Eaves and Saigal. J_3 and J_4 have exponentially decreasing grid size so are comparable while J_4 has mesh about twice that of K_2 . The pivot rules of Table 2 and the description of J_3 and J_4 using (5.1), (5.3) or (5.6) seem simpler than those for K_2 or K_3 .

A rudimentary Fortran program has been written for J_3 - it can be obtained from the author. No comparisons with K_3 have been made. A program for J_1 or J_2 should be easy to obtain from Table 1, and for J_4 a simple modification of the program for J_3 suffices. Note that, for $n = 1$, J_3 and K_3, and J_4 and K_2 coincide.

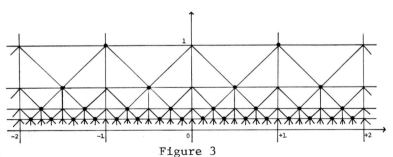

Figure 3

J_3^{0c} - heavy dots

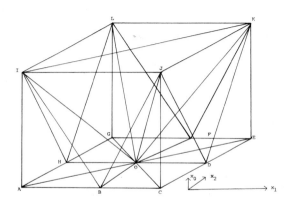

Figure 4

0   and   K   are in J_3^{0c} Simplices of J_3 containing

$0 = (\frac{1}{2}, \frac{1}{2}, \frac{1}{2})$, $(\upsilon(0) = (-1,-1,-1))$

π	s	(say $s_{\pi(j)} \cdots s_{\pi(n)}$ are set equal
⟨0,B,A,I⟩	(2,1,0)	(1,-1,-1) to +1)
⟨0,B,J,I⟩	(2,0,1)	(1,1,-1)
⟨0,B,C,J⟩	(2,1,0)	(1,1,-1)
⟨0,D,C,J⟩	(1,2,0)	(1,1,-1)
⟨0,D,K,J⟩	(1,0,2)	(1,1,1)
⟨0,D,E,K⟩	(1,2,0)	(1,1,1)
⟨0,F,E,K⟩	(2,1,0)	(1,1,1)
⟨0,F,K,L⟩	(2,0,1)	(1,1,1)
⟨0,F,G,L⟩	(2,1,0)	(1,-1,1)
⟨0,H,G,L⟩	(1,2,0)	(1,-1,1)
⟨0,H,L,I⟩	(1,0,2)	(1,-1,1)
⟨0,H,A,I⟩	(1,2,0)	(1,-1,-1)
⟨0,K,J,I⟩	(0,2,1)	(1,1,1)
⟨0,K,L,I⟩	(0,1,2)	(1,1,)

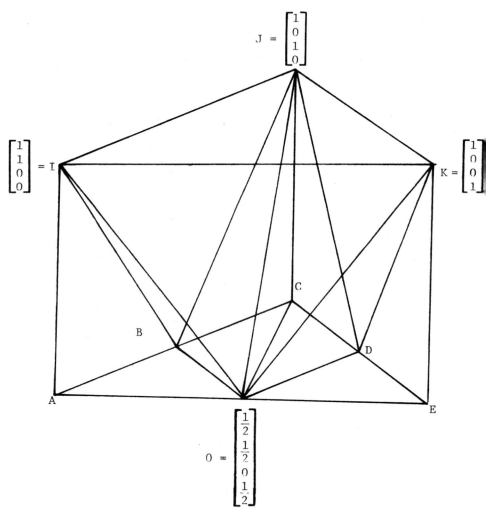

Simplices of J_4 are as in figure 4 with equivale
labelling.

Figure 5

This research was supported by National Science Foundation Grant GK-42092. The author would also like to thank Dr. Romesh Saigal for a suggestion which simplified the description of the triangulations considerably.

REFERENCES

[1]   Eaves, B. C., "Computing Kakutani Fixed Points," SIAM
      Journal on Applied Mathematics 21, 2 (1971) 236-244.

[2]   Eaves, B. C., "Homotopies for Computation of Fixed
      Points", Mathematical Programming 3 (1972), 1-22.

[3]   Eaves, B. C., and Saigal, R. "Homotopies for Computa-
      tion of Fixed Points on Unbounded Regions," Mathemati-
      cal Programming 3 (1972), 225-237.

[4]   Freudenthal, H. "Simplizialzerlegungen von Beschränk-
      ter Flachheit," Annals of Mathematics 43, 3(1942),
      580-582.

[5]   Hansen, T., "On the Approximation of a Competitive
      Equilibrium," Doctoral Thesis, Yale University, New
      Haven, Connecticut, 1968.

[6]   Kuhn, H. W. "Some Combinatorial Lemmas in Topology,"
      IBM J. Res. Develop. 4 (1960), 508-524.

[7]   Kuhn, H. W. "Simplicial Approximation of Fixed Points",
      Proc. Nat. Acad. Sci. 61 (1968), 1238-1242.

[8]   Kuhn, H. W. "Finding Roots of Polynomials by Systema-
      tic Search," to appear in Pivoting and Extensions,
      Mathematical Programming Studies 1.

[9]   Lefschetz, S. Introduction to Topology, Princeton
      University Press, Princeton, New Jersey, 1949.

[10] Shapley, L. S. "On Balanced Games without Side Pay-
      ments", in Mathematical Programming, T. C. Hu and
      Stephen M. Robinson, eds., Academic Press, N.Y., 1973.

[11] Whitney, H. Geometric Integration Theory, Princeton
      University Press, Princeton, New Jersey, 1957.

Fixed Point Iterations Using Infinite Matrices, III

B. E. Rhoades

ABSTRACT

Let E be a closed, bounded, convex subset of a Banach space X, $f: E \to E$. Consider the iteration scheme defined by $\bar{x}_o = x_o \in E$, $\bar{x}_{n+1} = f(x_n)$, $x_n = \sum_{k=0}^{n} a_{nk} \bar{x}_k$, $n \geq 1$, where A is a regular weighted mean matrix. For particular spaces X and functions f this iterative scheme converges to a fixed point of f.

[5] extends or generalizes related work of Browder and Petryshyn, Dotson, Franks and Marzec, Johnson, Kannan, Mann, and Reinermann. This paper continues investigations begun in [6].

Let f be a continuous self-mapping of [0.1] . We
shall be concerned with the iteration scheme

$$x_o \in [0,1] \tag{1}$$

$$x_{n+1} = (1-c_n)x_n + c_n f(x_n) , \quad n \geq 0 , \tag{2}$$

where $\{c_n\}$ is a real sequence satisfying (i) $c_o = 1$,

(ii) $0 \leq c_n \leq 1$ for all $n > 0$, (iii) $\sum c_n$ diverges, and

(iv) $\lim_n c_n = 0$.

Equation (2) can be cast in the following form:

$$x_{n+1} = \sum_{k=0}^{n} a_{nk} f(x_k) , \quad \text{where} A = (a_{nk}) \text{ is the triangular}$$

matrix defined by

$$a_{nn} = c_n$$

$$a_{nk} = c_k \prod_{j=k+1}^{n} (1-c_j) \quad \text{for } k < n \tag{3}$$

$$a_{nk} = 0 \quad \text{for } k > n .$$

Conditions (i) - (iii) are sufficient to guarantee that A
is a regular method of summability.

The iteration scheme has recently been investigated independently by Dotson [2], Outlaw and Groetsch [3], and
Reinermann [4]. A theorem equivalent to the following appears in [5].

Theorem R1 Let f be a continuous self-mapping of [0,1]
with $\{c_n\}$ satisfying (i) - (iv). Then (1), (2) converges
to a fixed point of f .

In order to extend Theorem R1 to more general spaces, one must replace the continuity of f by some growth condition. Fixed point theorems which apply (1), (2) to operators satisfying various growth conditions appear in [5].

We include here one result on multivalued monotone operators in Hilbert space.

<u>Theorem 1</u> Let $\{c_n\}$ be a real positive monotone decreasing sequence satisfying $(1+\sigma)^{-1} \geq c_n \geq c_{n-1} \big/ (1+c_{n-1})$, $n > 1$.

Let A be defined by (3), T a multivalued monotone operator with open domain $D(T)$ in a Hilbert space H , and $y \in R(I+T)$. Then there exists a neighborhood $N \subset D(T)$ of $\bar{x} \in (I+T)^{-1} y$ and a real positive number σ_1 , such that, for any $\sigma \geq \sigma_1$, any initial guess $x_1 \in N$, and any single valued section T_o of T , the sequence $\{x_n\}$ defined by

$$x_{n+1} = (1-c_n)x_n + c_n(y-T_o x_n) \ , \quad n \geq 1$$

remains in $D(T)$ and converges to $\bar{x}$ with extimate $\| x_n - \bar{x} \| = 0(c_n^{1/2})$. The sequence $\{x_n + T_o x_n\}$ is A-summable to y .

The special case of the above theorem for $c_n = (n+\sigma)^{-1}$ appears in [1]. The reader should consult [1] for special definitions and terminology.

The convergence of $\{x_n\}$ is proved exactly as that in [1] by observing that the conditions on $\{c_n\}$ guarantee that $(1-c_n)^2 d_n^2 + c_n^2 \leq d_{n+1}^2$, where $d_n^2 = c_{n-1}$. To show the A-summability of $\{x_n + T_o x_n\}$, use an induction argument to verify that $x_{n+1} = x_1 \prod_{j=1}^{n} (1-c_j) + \sum_{k=1}^{n} a_{nk} (y-T_o x_k)$. The

condition $c_n \geq c_{n-1}/(1+c_{n-1})$ implies $1 - c_n \leq (1+c_{n-1})^{-1}$

and $(1 + c_{n-1})^{-1} \leq c_n/c_{n-1}$. Therefore $\prod_{j=2}^{n} (1-c_j) \leq$

$c_n/c_1 \to 0$, and A is regular.

The monotonicity of $\{c_n\}$ and the condition $c_n \geq$ $c_{n-1}/(1+c_{n-1})$ are necessary in order to keep $\{x_n\}$ in $D(T)$.

The author has been informed, through private correspondence, that the condition that T_o be a single valued section can be replaced by the condition that for each $n \geq 1$, $x_{n+1} \in (1-c_n)x_n + c_n(y-Tx_n)$.

For the remainder of this note we shall restrict our attention to continuous self-mappings of $[0,1]$. For completeness we quote the following from [6], where

$M = \sup\{x \in [0,1] | f(x)=x\}$ and $m = \inf \{ x \in [0,1] | f(x) = x\}$.

__Theorem R2__   Let f be a continuous, nondecreasing self-mapping of $[0,1]$. If $\{c_n\}$ satisfies (i) - (iii), then (1), (2) converges to a fixed point of f .

__Theorem R3__   Let f be a continuous, nondecreasing self-mapping of $[0,1]$, $c_o = d_o = 1$, $0 \leq c_n \leq d_n \leq 1$ for all $n > 0$, $\sum c_n$ diverges, $x_o = y_o$, and for $n > 0$, $\{x_{n+1}\}$ satisfies (2) and $\{y_{n+1}\}$ satisfies $y_{n+1} = (1-d_n)y_n + d_n f(y_n)$.

If $x_o > M$, then $x_n \geq y_n$ for all n . If $x_o < m$, then $x_n \leq y_n$ for all n . If there exists a pair of

distinct adjacent fixed points p,q satisfying $m \leq p < q \leq$
M , and $x_0 \in (p,q)$, then $f(x) > x$ for $x \in (p,q)$ implies $x_n \leq y_n$, and $f(x) < x$ for $x \in (p,q)$ implies
$x_n \geq y_n$.

A consequence of Theorem R3 is that, for nondecreasing continuous functions, Picard iteration is the best procedure in the sense that $|f^n(x_0) - p| \leq |x_n - p|$, for each n , where p is the fixed point of f to which (1), (2) converges.

We add one more fact for increasing functions.

Theorem 2 Let f , $\{c_n\}$ satisfy the conditions of Theorem R2. If $y_0 > x_0$, then $y_{n+1} \geq x_{n+1}$ for each n , where
$y_{n+1} = (1-c_n)y_n + c_n f(y_n)$ and x_{n+1} satisfies (2).
The proof is immediate.

An analysis of the proof of Theorem R2 shows that the initial choice of x_0 determines to which fixed point of f
(2) will converge. For example, suppose f has three fixed points p,q,r , with $0 < p < q < r < 1$. If $x_0 \in [0,q)$, then $x_n \to p$, whereas $x_0 \in (q,1]$ implies $x_n \to r$.
Theorem 3 demonstrates that, for such an f , the closer x_0
is to p or r , the better the iteration scheme will be.

It can be shown by examples that there are no properties comparable to Theorems 2 and R3 for decreasing functions. However, one can obtain an error estimate for nonincreasing functions, which we now do.

Theorem 3 Let f be a continuous nonincreasing self-mapping of [0,1]. For each n such that $2c_n \leq 1$, we have

$|x_{n+1}-p| \leq |x_{n+1}-f(x_n)|$, where p is the unique fixed

point of f .

The existence of a unique p is obvious. From (2),

$x_{n+1} - p = (1-c_n)(x_n-p) + c_n(f(x_n)-p)$. If $x_n \geq p$, then

$f(x_n) \leq p$ and $x_{n+1}-p \leq (1-c_n)(x_n-p) \leq (1-c_n)(x_n-f(x_n))$.

Also $x_{n+1} - p \geq c_n(f(x_n)-p) \geq c_n(f(x_n)-x_n)$. Note that

$(1-c_n)(x_n-f(x_n)) + c_n(f(x_n) - x_n) = (1-2c_n)(x_n-f(x_n)) \geq 0$,

so that $|x_{n+1}-p| \leq (1-c_n)|x_n - f(x_n)|$.

If $x_n < p$, then a similar argument shows again that

$|x_{n+1}-p| \leq |(1-c_n)|x_n - f(x_n)|$.

From (2) we have, $x_{n+1}-f(x_n) = (1-c_n)x_n + (c_n-1)f(x_n) =$

$(1-c_n)(x_n-f(x_n))$, and the result follows.

No error estimate for increasing functions has been

found, although it can be shown that $|x_{n+1}-p| > \max \{(1-c_n) \cdot$

$|x_{n+1}-x_n|$, $c_n|f(x_n) - x_n|$, $c_n|f(x_n) - x_{n+1}|$, $(1-c_n) \cdot$

$|f(x_n) - x_{n+1}|\}$, and there exist functions such that

$|x_{n+1}-p| > |x_n - x_{n+1}|$.

A sometimes useful ad hoc criterion is the error esti-

mate $\varepsilon_n = |f(x_n) - x_n|$.

The identity matrix, applied to the function f(x) = 1-x,

shows that condition (iv) in Theorem R1 is necessary. Thus,

the restriction, in Theorem 3, that $2c_n \leq 1$, is a mild one.

In Theorem 3, the condition that f be nonincreasing is

necessary. For, if f is monotone increasing over [p,1],

where p is the largest fixed point of f , then $x_n > p$

implies $f(x_n) > p$, so that $x_{n+1}-p > x_{n+1}-f(x_n) = (1-c_n) \cdot$
$(x_n-f(x_n)) \geq 0$.

In Theorem R1, condition (iii) on $\{c_n\}$ is necessary
to ensure that A be regular. We shall now examine the behavior of the iteration scheme (1), (2) under the assumption

(v) : $\sum c_n$ converges.

__Theorem 4__ Let A be defined by (3) with $\{c_n\}$ satisfying
(i), (ii), and (v). Then A is coercive.

A coercive matrix is one which sums all bounded sequences; i.e., it is a compact operator over c , the space
of convergent sequences. The conditions on $\{c_n\}$ imply that

A is conservative, but not regular. For any conservative
matrix A , and any convergent sequence z , $\lim_{A} z =$
$\chi(A) \lim z + \sum_k a_k z_k$, where $\chi(A) = t - \sum_k a_k$, $t =$
$\lim_n \sum_k a_{nk}$, and $a_k = \lim_n a_{nk}$. (See, e.g., [7, p. 93].)

From (v), $\lim_n c_n = 0$. For simplicity we shall assume that each $c_n < 1$ for $n > 0$. Then, from [5], A
can be considered as a weighted mean matrix of the form

$$a_{nk} = P_k/P_n \quad , \text{ where } \quad P_0 > 0 \quad , \quad P_k = P_0 c_k \Big/ \prod_{j=1}^{k} (1-c_j), \quad k > 0,$$

$$P_0 = P_0 \quad , \quad P_n = \sum_{k=0}^{n} P_k = P_0 \Big/ \prod_{j=1}^{n} (1-c_j), \quad n > 0 \quad . \quad (v) \text{ im-}$$

plies that the product converges. Let $P = P_0 \Big/ \lim_n \prod_{j=1}^{n} (1-c_j)$.
Then $a_k = P_k/P$, and $\chi(A) = 1 - 1 = 0$.

A matrix A is coercive if and only if

$$\lim_{n} \sum_{k=0}^{\infty} |a_{nk} - a_{k}| = 0 \ . \tag{4}$$

In our case $a_{nk} - a_{k} = p_{k}/P_{n} - p_{k}/P \geq 0$, so that the left hand side of (4) reduces to $\chi(A)$, which is zero.

Let f be any continuous function defined on [0,1] . Then $\{f(x_{n})\}$ is bounded. Since A is coercive, $\lim_{n} x_{n+1}$ exists, i.e., $\{x_{n}\}$ is convergent. Thus $\{f(x_{n})\}$ is con-vergent, and $\lim_{n} x_{n} = \sum_{k=0}^{\infty} a_{k}f(x_{k})$.

If any of c_{k}'s is 1 then, from (v), there can be only a finite number of them. Let $N > 0$ denote the largest val-ue of k such that $c_{k} = 1$. From (3), $a_{nk} = 0$ for $k <$ $N \leq n$, and $a_{k} = 0$ for each such k . Changing the en-tries in a finite number of columns does not affect the size of the convergence domain of any conservative matrix, so that A is still coercive. In any case, $\lim_{n} x_{n} = \sum_{k=N}^{\infty} a_{k}f(x_{k})$, where N is the largest value of k for which $c_{k} = 1$.

We did not assume f to be a self-mapping, so it need not have any fixed points in [0,1]. Even if f is a self-mapping, $\{x_{n}\}$ need not converge to a fixed point of f . For example, let $f(x) = (1+x)/2$, $c_{n} = 2^{-n}$, $n \geq 0$, and choose $x_{o} = 0$. Then $x_{1} = 1/2$, and, from (2), we have $x_{n+1} - x_{n} = c_{n}(1-x_{n})/2$, so that $\sum_{k=1}^{n} (x_{k+1}-x_{k}) =$

$$\sum_{k=1}^{n} c_k(1-x_k)/2 \quad , \text{ or } \quad x_{n+1} = x_1 + \left[\sum_{k=1}^{n} c_k - \sum_{k=1}^{n} c_k x_k\right] \Bigg/ 2 \ .$$

From the proof of Theorem R1, $x_n \uparrow$ in n so that $x_k \geq 1/2$,

$k \geq 1$, $x_{n+1} \leq 1/2 + \sum_{k=1}^{n} (2^{-k} - 2^{-k-1})/2$, and $\lim_n x_n \leq$

$3/4 < 1$.

To gain some inductive evidence on the behavior of the iteration scheme (1), (2) for monotone decreasing functions, a program was written to approximate the fixed points of the functions $f(x) = 1 - x^m$, $g(x) = (1-x)^m$ for $1 \leq m \leq 6$, with $c_n = [(n+1)(n+2)]^{-1/k}$, $3 \leq k \leq 8$. The fixed point of each function was first found by the bisection method, accurate to 10 places. Then both the iteration scheme (1), (2) and Newton-Raphson were employed to find each fixed point to within 8 places, using initial guesses of $x_0 = .1, .2, \ldots$, .9 . An examination of the output allows one to make the following observations:

    1. Newton-Raphson converges faster than (1), (2) . This is not surprising, since (1), (2) converges linearly, whereas N - R is a quadratic method. However, whereas N - R converges more rapidly for x_0 near the fixed point, (1), (2) appears to converge somewhat independently of the initial guess. For example, with $m = 4$ or 6, $k = 4$, (1), (2) converges to the fixed point of f in exactly 8 iterations, for each choice of x_0 .

    2. The most efficient choice of k is 5 , for $m < 3$, and 4 for $m \geq 3$. The number of iterations required increases with the distance from k to 4 or 5 . For example, a separate program was run for f with $m =$

2,3,4, k = 2 and x_o = .9 . In each case (1), (2) was only accurate to 5 places after 400 iterations! Consequently (C,1) summability should not be used in (2) for finding fixed points.

   3.   The program also computed the Picard iterates of f and g . In each case, if the initial guess exceeded the fixed point then the even terms of $\{f^n(x_o)\}$ increased monotonely to 1, and the odd terms decreased monotonely to zero. If x_o was smaller than the fixed point, the behavior of the even and odd terms was interchanged.

   4. $\{x_n - p\}$ need not be an alternating sequence.

   I take this opportunity to thank Douglas A. Ford for programming the problems.

   More numerical work is required to gain additional insight into the iterative scheme. The following are interesting and important unresolved questions:

   1.   Does there exist an error estimate comparable to Theorem 3 for increasing functions?

   2.   Is there an optimal choice of $\{c_n\}$ for decreasing functions?

   3.   Does Theorem R1 remain true if [0,1] is replaced by the unit square, and no additional restrictions are placed on f ?

REFERENCES

[1] Bruck, R. E., Jr., "The Iterative Solution of the Equation y∈x+Tx for a Monotone Operator T in Hilbert Space", Bull. Amer. Math. Soc. 79 (1973), 1258-1261.

[2] Dotson, W. G., Jr., "On the Mann Iterative Process", Trans. Amer. Math. Soc. 149 (1970), 65-73.

[3] Outlaw, C. L., and Groetsch, C. W., "Averaging Iteration in a Banach Space", Bull. Amer. Math. Soc. 75 (1969), 430-432.

[4] Reinermann, J., "Über Toeplitzsche Iterationsverfahren und Einige Ihre Anwendungen in der Konstruktiven Fixpunktheorie", Studia Math. 32 (1969), 209-227.

[5] Rhoades, B. E., "Fixed Point Iterations Using Infinite Matrices", to appear in Trans. Amer. Math. Soc.194(1974).

[6] Rhoades, B. E., "Fixed Point Iterations Using Infinite Matrices, II", to appear in Constructive and Computational Methods for Differential and Integral Equations, Springer-Verlag Lecture Notes Series.

[7] Wilansky, A., Functional Analysis, Blaisdell, 1964.

Some Aspects of Mann's Iterative Method
for Approximating Fixed Points

C. W. Groetsch

ABSTRACT

    Some connections between summability theory and Mann's
iterative process for affine mappings are exhibited and some
questions are posed. Results are given for linear operator
equations of the first and second kind using a special case
of Mann's process and an iterative method for generalized
solutions of singular linear operator equations of the first
kind is studied.

1.   Introduction

In 1953 W. R. Mann [17] published a general iterative method for approximating fixed points of a mapping T de-fined on a linear space E . For a given infinite, lower triangular, regular row-stochastic matrix $[a_{nk}]$ Mann gen-erated sequences $\{x_n\}$ and $\{v_n\}$ in E by the formulas

$$v_n = \sum_{k=0}^{n} a_{nk} x_k \quad , \quad x_{n+1} = Tv_n \qquad (1.1)$$

where $x_o \in E$ is arbitrary. For example, if $[a_{nk}]$ is the infinite identity matrix, then each of the sequences $\{x_n\}$ and $\{v_n\}$ is the sequence $\{T^n x_o\}$ of Picard iterates of the mapping T . Mann showed that if either of the sequences $\{x_n\}$ or $\{v_n\}$ converges, then the other also converges to the same point and their common limit is a fixed point of T . The value of the process (1.1) resides in the fact that under certain conditions it converges even when the sequence of Picard iterates does not. For example, if $[a_{nk}]$ is the Cesaro matrix then the method (1.1) converges to a fixed point of a continuous mapping T of a closed interval into itself [17, 5].

In this paper we will survey some aspects of this it-erative method when the mapping T is affine. Linear opera-tor equations of the first and second kind may be considered within the framework of fixed point problems for affine

mappings and in particular the results are applicable in the
theory of linear integral equations. In the next section,
the connection between Mann's process for affine mappings and
summability theory is investigated and some questions are
posed. Section 3 deals with a special case of Mann's method
and linear operator equations of the first and second kinds.
The final section of this paper is concerned with singular
operator equations of the first kind and iterative methods
for the generalized inverse. In this section a strong inter-
play between summability theory and the iterative method is
apparent.

2. Affine Mappings

If the mapping T is affine then since $A = [a_{nk}]$ is
row-stochastic it follows that the sequence (1.1) may be
written as

$$v_n = \sum_{k=0}^{n} c_{nk} \, T^k \, x_o \qquad (2.1)$$

where the matrix $C = [c_{nk}]$ depends only on A . Thus the
sequence $\{v_n\}$ may be regarded as the transform of the Picard sequence $\{T^n x_o\}$ by the infinite matrix C . It is
known that if A is regular then so is C [19] and hence
the Mann process applied to an affine mapping generates a
sequence which is a summability transform of the sequence of
Picard iterates of the mapping.

Given two summability matrices G and H one says
that G is __stronger__ than H , denoted $G \geq H$, if every
sequence which is summed by H is also summed by G to the

same value. If $G \geq H$ and $H \geq G$, then G and H are called equivalent, denoted $G \approx H$. Finally if $G \geq H$ and G is not equivalent to H , then G is said to be strictly stronger than H which is denoted $G > H$. Rhoades [23] has pointed out the difficulty in using comparison theorems from summability theory in studying the Mann process for general mappings T . However, if T is affine, then it would seem that the connections between the representations (1.1) and (2.1) would be amenable to study by summability techniques. A natural conjecture is that if A and A' are two matrices in the scheme (1.1) with $A \geq A'$ and if C and C' are the corresponding matrices in the representation (2.1) then $C \geq C'$. We will show below that this is not the case and pose a similar question.

The Mann process has been most extensively investigated for the class of matrices which are called segmenting in [19] and [7] (see [15], [24], [3], [22], [12], [10], [21], and [23]). These are the matrices which satisfy $a_{oo} = 1$, $a_{nn} = \lambda_n$ for $n > 0$, where $0 < \lambda_n \leq 1$ and $a_{n+1,k} = (1-\lambda_{n+1})a_{nk}$ for $k \leq n$. In this case the iteration (1.1) becomes

$$v_{n+1} = (1-\lambda_{n+1})v_n + \lambda_{n+1}Tv_n . \qquad (2.2)$$

A segmenting matrix is clearly determined by the diagonal entries $\{\lambda_n\}$ and the matrix determined by the above conditions will be denoted by $A(\lambda_n)$. The matrix associated with $A(\lambda_n)$ by way of (2.1) will be denoted by $C(\lambda_n)$. Note that $A(\frac{1}{n+1})$ is the Cesaro (C,1)-matrix and $C(\frac{1}{n+1})$ is the Lototsky matrix [11]. Also if $\lambda_n = \lambda$ for all n then $C(\lambda)$ is the Euler-Knopp matrix with parameter λ .

In general $C(\lambda_n)$ is a generalized Lototsky matrix [11, 6].

It is easy to show that the matrix $A(\lambda_n)$ is regular, if and

only if, $\sum \lambda_n = \infty$ [3], and that the inverse of $A(\lambda_n)$ is

regular, if and only if, $\underline{\lim} \lambda_n > 0$. In such a case it is

clear that $A(\lambda_n)$ sums only convergent sequences.

Proposition 2.1 $A(\lambda_n)$ is equivalent to convergence, if and

only if, $\underline{\lim} \lambda_n > 0$.

If we take $0 < \lambda < \mu \le 1$ then $A(\lambda) \approx A(\mu)$ by this proposition but $C(\lambda) > C(\mu)$ since the strength of the Euler-
Knopp methods increases with a decrease in the parameter
[27]. It follows that the transition from the matrix A in
(1.1) to the matrix C in (2.1) is not monotone with respect
to the partial ordering $\ge$. By considering the regularity
of the product method $A(\lambda_n)A^{-1}(\mu_n)$ one can easily prove the
following inclusion theorem for segmenting matrices.

Proposition 2.2 If $\sum \lambda_n = \infty$ and $\lambda_{n+1}\mu_{n+1}^{-1} \le \lambda_n\mu_n^{-1} \le 1$ for

$n \ge n_o$, then $A(\lambda_n) \ge A(\mu_n)$. If in addition $\underline{\lim} \lambda_n\mu_n^{-1}$

$= 0$, then $A(\lambda_n) > A(\mu_n)$.

This proposition along with various examples of generalized
Lototsky matrices indicate an affirmative answer to the following question, at least for segmenting matrices.

Question 2.1 If $A > A'$ in (1.1) does it follow that
$C > C'$ in (2.1)?

Consider the Mann process (2.2) with $\lambda_n = \lambda$ for all n (or

equivalently (1.1) with $A = A(\lambda)$) and assume that the

process converges. Schaefer [24] has shown that in general
the fixed point $p(\lambda)$ to which the method converges indeed
depends upon the parameter λ . However, if T is affine
then the sequence (1.1) using the matrix $A(\lambda)$ is the Euler-
Knopp transform with parameter λ of the sequence $\{T^n x_0\}$.
Since the Euler-Knopp methods are consistent [27] (this fol-
lows since $0 < \lambda \leq \mu \leq 1$ implies $E(\lambda) \geq E(\mu)$) we have the
following result.

Proposition 2.3   If T is affine then the fixed point (if
any) to which (1.1) with $A = A(\lambda)$ converges is independent
of λ .

If T is quasi-nonexpansive [3] (i.e., $\| Tx-p \| \leq \| x - p \|$
for each fixed point p) and if p is a fixed point of T,
then the functions $f_n(\lambda) = \| v_n(\lambda) - p \|$ are continuous
and decreasing (where $v_n(\lambda)$ denotes the iterate generated
by (2.2) using the parameter λ). An application of Dini's
theorem then gives

Proposition 2.4   If T is affine and quasi-nonexpansive and
if (1.1) with $A = A(\lambda)$ converges to a fixed point p , then
the convergence is uniform for λ in compact subsets of
(0,1) .

In order to apply Dini's theorem above it is only necessary
that $\lambda \to p(\lambda)$ be continuous; hence the following question
seems relevant.

Question 2.2   Suppose T is quasi-nonexpansive (not necessarily affine) and (1.1) with $A = A(\lambda)$ converges to a fixed
point $p(\lambda)$, is the mapping $\lambda \to p(\lambda)$ continuous?

If under certain circumstances this question could be answered in the affirmative then a result similar to Proposition 2.4 could be given for non-affine mappings T .

3.   Linear Operator Equations

In this section we shall discuss some iterative methods for linear operator equations of the first and second
kinds. A linear operator equation of the second kind has
the form

$$u - Tu = f \qquad (3.1)$$

where T is a linear operator on a Banach space E and f
is a fixed element of E . Note that any solution of (3.1)
is a fixed point of the affine operator T_f defined by

$T_f x = Tx + f$ and hence the Mann method may be applied to

T_f in attempting to approximate solutions of (3.1). In

particular if we apply the segmenting Mann process (2.2)
with parameters $\{\lambda_n\}$ where $0 \leq \lambda_n \leq 1$ to the affine map-

ping T_f we have the iterative process

$$v_{n+1} = (1-\lambda_{n+1})v_n + \lambda_{n+1}(Tv_n+f) \qquad . \qquad (3.2)$$

In [8] the author studied this process from the point of view
of Dotson's elegant theory of linear iteration based on abstract ergodic theory [2]. If the operators $A_n(T)$ $(n \geq 0)$
are defined by

$$A_n(T) = \prod_{j=1}^{n} [(1-\lambda_j)I + \lambda_j T] \qquad (3.3)$$

and $B_n(T)$ are defined by $B_0(T) = 0$ and

$$B_{n+1}(T) = ((1-\lambda_{n+1})I + \lambda_{n+1}T)B_n(T) + \lambda_{n+1}I \qquad (3.4)$$

then in order to apply Dotson's general result it must be shown that $\{A_n(T)\}$ is a system of almost invariant integrals for the semigroup $G(T) = \{I,T,T^2,...\}$ and that $\{B_n(T)\}$ is a system of "companion integrals" (see [2]). If the operator T is asymptotically bounded (i.e., $\|T^n\| \leq M$), then the operators $\{A_n(T)\}$ would be uniform almost invariant integrals for $G(T)$ if the matrix $C(\lambda_n)$ is <u>strongly regular</u>, that is

$$\lim_n \sum_{k=0}^{n-1} |c_{n,k+1} - c_{nk}| = 0 \quad .$$

The matrix $C(\lambda_n) = [c_{nk}]$ can be interpreted in terms of probability theory in the following way. If we consider a sequence of independent random trials in which the probability of success on the ith trial is λ_i and let X_i be the random variable which indicates the outcome of the ith trial, i.e., $P(X_i=1) = \lambda_i$ and $P(X_i=0) = 1 - \lambda_i$, then $P(S_n=k) = c_{nk}$ where $S_n = X_1 + \cdots + X_n$. If $var(S_n) = \sum_{i=1}^{n} \lambda_i(1-\lambda_i) \to \infty$ then the DeMoivre-Laplace theorem can be applied to show that the matrix $C(\lambda_n)$ is strongly regular (see [6] and [8]) and hence that $\{A_n(T)\}$ forms a system of uniform almost invariant integrals for $G(T)$. An application of Dotson's theorem then gives

Proposition 3.1 Let T be an asymptotically bounded linear operator defined on a reflexive Banach space E and suppose $0 \leq \lambda_n \leq 1$ and $\sum_n \lambda_n (1-\lambda_n) = \infty$. (a) If $f \in R(I-T)$, then $\{v_n\}$ defined by (3.2) converges to a solution u of (3.1). (b) If $\{v_n\}$ clusters weakly at $y \in E$, then y is a solution of (3.1) and $\{v_n\}$ converges strongly to y .

If the context is specialized to self-adjoint linear operators in Hilbert space we obtain

Proposition 3.2 In addition to the hypotheses above, assume that E is a Hilbert space and T is self-adjoint. (a) If $f \in E$, then $\{v_n\}$ is a minimizing sequence for equation (3.1) in the sense that

$$\lim_n \| v_n - Tv_n - f \| = \inf \{\| v-Tv-f \| : v \in E\} .$$

(b) If $f \in R(I-T)$, then $\{v_n\}$ converges to the solution of (3.1) which is nearest to v_0 .

Recently Lardy [6] considered iteratively generated sequences of operators of the type (3.3). Among the results he established is the following.

Proposition 3.3 [16]. Let T be a bounded linear operator on a Banach space E . Suppose the spectrum of T satisfies $\sigma(T) \subset \{z \mid |\lambda z + (1-\lambda)| \leq 1\}$ where $\lambda > 0$ and if $1 \in \sigma(T)$ then 1 is a pole of order one of the resolvent operator $R(z;T)$. Suppose further that $\{\lambda_n\}$ satisfies $0 < \lambda_n$, $\lim \lambda_n < \lambda$ and $\sum_n \lambda_n (1-\lambda_n) = \infty$. Then the operators $\{A_n(T)\}$ defined by (3.3) converge in the uniform operator topology to P_N , the spectral projection of E

onto N(I-T) .

If the operator I - T is invertible, i.e., $1 \notin \sigma(T)$, then Lardy obtains as a corollary that the sequence of operators $\{V_n\}$ defined by

$$V_{n+1} = (1-\lambda_{n+1})V_n + \lambda_{n+1}(TV_n+I)$$

converges in the uniform operator topology to $(I-T)^{-1}$. Related results based on abstract ergodic theory for more general processes have been given by Koliha [14].

A linear operator equation of the <u>first</u> kind has the form

$$Ax = f \qquad\qquad (3.5)$$

where A is a linear operator on E and $f \in E$. By set-ting A = I - T (3.5) is transformed into an equation of the second kind and vice versa. The iteration (3.2) becomes for operator equations of the first kind

$$v_{n+1} = v_n + \lambda_{n+1}(f-Av_n) \quad . \qquad\qquad (3.6)$$

This iteration bears a formal resemblance to various gradient methods for linear operator equations, in particular the method of steepest descent [18].

<u>Proposition 3.4</u> Let A be a nonnegative (i.e., $\langle Ax,x\rangle \geq 0$) bounded linear operator on a Hilbert space E . Suppose $0 < \lambda_n \leq 2\|A\|^{-1}$ and $\sum\lambda_n(2\|A\|^{-1}-\lambda_n) = \infty$, and suppose $\{v_n\}$ is defined by (3.6). (a) For each $f \in E$, $\lim \|Av_n-f\| = \inf \{\|Av-f\| \mid v \in E\}$. (b) If $f \in R(A)$ then $\{v_n\}$ converges strongly to the solution of (3.5) which is

nearest to v_o . (c) If $\{v_n\}$ clusters weakly at $w \in E$, then w is a solution of (3.5) and $\{v_n\}$ converges strongly to w .

4. Singular Equations

We will now investigate operator equations of the first kind

$$Ax = f \qquad\qquad (4.1)$$

where A is a bounded linear operator from a Hilbert space H_1 into a Hilbert space H_2 ($A \in L(H_1,H_2)$) with closed range $R(A)$. Equation (4.1) is of course singular if $f \notin R(A)$ but even in this case one can speak of "solving" the equation if the notion of solution is suitably extended. A natural way to extend the idea of solubility is to consider least squares solutions of (4.1). An element $u \in H_1$ is called a least squares solution of (4.1) if

$$\| Au - f \| = \inf \{ \| Ax - f \| : x \in H_1 \} \ .$$

It is easy to show that the set of least squares solutions of (4.1) coincides with the set of solutions of

$$A^*Au = A^*f \qquad\qquad (4.2)$$

and that there is a unique least squares solution of smallest norm. The operator which assigns to each $f \in H_2$ the solution of (4.2) with minimal norm is called the generalized inverse of A and is denoted by A^+ . M. Z. Nashed [18] has written a valuable survey article on generalized inverses which contains a comprehensive bibliography.

Desoer and Whalen [1] have given a particularly useful characterization of the generalized inverse of a bounded linear operator A with closed range as the unique operator A^+ in $L(H_2,H_1)$ which satisfies

$$A^+Ax = x \qquad \text{for} \quad x \in R(A^*)$$
$$A^+y = 0 \qquad \text{for} \quad y \in N(A^*) \quad . \tag{4.3}$$

Quite recently the author used this characterization along with some notions from summability theory and spectral theory to give a general representation theorem for the generalized inverse. A crucial component in the proof is the fact that if R(A) is closed then the restriction of A^*A to H = $R(A^*)$ is bounded below. It follows that if $\tilde{A}$ denotes the operator in L(H,H) which is the restriction of $I - A^*A$ to H , where I is the identity operator on H_1 , then

$\sigma(\tilde{A}) \subset (-\infty,1)$. The spectral theorem for self-adjoint operators and the characterization (4.3) can then be used to prove the following theorem.

Proposition 4.1 Suppose Ω is an open set with $\sigma(\tilde{A}) \subset \Omega \subset$ $(-\infty,1)$ and let $\{S_t(x)\}$ be a net of continuous real functions on Ω such that $\lim S_t(x) = 1/(1-x)$ uniformly on $\sigma(\tilde{A})$, then $A^+ = \lim S_t(\tilde{A})A^*$, where the limit is in the uniform topology for $L(H_2,H_1)$.

By choosing for $S_t(x)$ various summability transforms of the geometric series $1 + x + x^2 + \cdots$ the author gave seven specific representations for A^+ in [9]. We will now give another representation which is based on generalized Lototsky summability. This representation gives an operator valued

iterative method of type (3.6) for computing A^+ . To ob-

tain the representation we use the specialization of a fact

proved by Jakimovski [11] that if

$$S_n(x) = \sum_{k=0}^{n} \lambda_{k+1} \prod_{j=1}^{k} [1-\lambda_j + \lambda_j x]$$

where

$$0 \leq \lambda_n \leq 1, \ \lambda_n \to 0, \ \sum \lambda_n = \infty \ \text{ and } \ \sum \lambda_n^2 (1-\lambda_n)^{-2} < \infty \quad (4.4)$$

then $S_n(x) \to 1/(1-x)$ uniformly on compact subsets of

Rex < 1 . If we set $A_{n+1} = S_n(A)A^*$ it is easy to see that

the sequence of operators $\{A_n\}$ is generated recursively by

the formula

$$A_{n+1} = A_n + \lambda_{n+1} (A^* - A^* A A_n) \quad , \quad (A_0 = 0) \quad . \qquad (4.5)$$

Proposition 4.1 then gives

Proposition 4.2 Suppose $A \in L(H_1, H_2)$ has closed range.

Let $\{\lambda_n\}$ be a sequence satisfying (4.4), then the opera-

tors generated by (4.5) converge in the uniform norm on

$L(H_2, H_1)$ to A^+ .

Note that if we set $v_n = A_n f$, then (4.5) produces a point-

wise iteration process of the form (3.6) for equation (4.2).

Petryshyn [20] and Showalter [25] have given iterative meth-

ods of type (4.5) which involve a fixed averaging parameter

which depends on $\| A \|^2$ and Lardy [16] has given a method

which uses a sequence of parameters which must satisfy a

condition involving $\|A\|^2$, however, in the method given above the sequence $\{\lambda_n\}$ is independent of any estimates on the size of $\|A\|$.

REFERENCES

[1] Desoer, C. A., and Whalen, B. H., "A Note on Pseudo-
     inverses", SIAM Journal of Applied Mathematics, 11
     (1963), 442-447.

[2] Dotson, W. G., Jr., "An Application of Ergodic Theory
     to the Solution of Linear Functional Equations in
     Banach Spaces", Bulletin of the American Mathematical
     Society, 75 (1969), 347-352.

[3] Dotson, W. G., Jr., "On the Mann Iterative Process",
     Trans. Amer. Math. Soc., 149 (1970), 65-73.

[4] Dotson, W. G., Jr., "On the Solution of Linear Func-
     tional Equations by Averaging Iteration", Proc. Amer.
     Math. Soc., 25 (1970), 504-506.

[5] Franks, R. L., and Marzec, R. P., "A Theorem on Mean
     Value Iterations", Proc. Amer. Math. Soc., 30 (1970),
     324-326.

[6] Groetsch, C. W., "Remarks on a Generalization of the
     Lototsky Summability Method", Boll. Un. Mat. Ital.,
     (series 4) 5 (1972), 277-288.

[7] Groetsch, C. W., "A Note on Segmenting Mann Iterates",
     J. Math. Anal. Appl. 40 (1972), 369-372.

[8] Groetsch, C. W., "Ergodic Theory and Iterative Solu-
     tion of Linear Operator Equations", Applicable Anal.
     (to appear).

[9] Groetsch, C. W., "Representations of the Generalized
     Inverse", J. Math. Anal, Appl., (to appear).

[10] Hillam, B. P., Ph.D. Thesis, University of California,
     Riverside, June 1973.

[11] Jakimovski, A., "A Generalization of the Lototsky Meth-
     od of Summability", Mich. Math. J., 6 (1959), 277-290.

[12] Johnson, G. G., "Fixed Points by Mean Value Iteration", Proc. Amer. Math. Soc., 34 (1972), 193-194.

[13] Kammerer, W., and Nashed, M. Z., "Iterative Methods for Best Approximate Solutions of Linear Integral Equations of the First and Second Kinds", J. Math. Anal. Appl., 4 (1972), 547-573.

[14] Koliha, J. J., "Ergodic Theory and Averaging Iterations", Can. J. Math., 25 (1973), 14-23.

[15] Krasnoselskii, M. A., "Two Remarks About the Method of Successive Approximation", Uspehi Mat. Nauk 10 (1955), 123-127 (Russian).

[16] Lardy, L. J., "Some Iterative Methods for Linear Operator Equations with Applications to Generalized Inverses", Technical Report TR 63-65, Department of Mathematics, University of Maryland, November 1973.

[17] Mann, W. R., "Mean Value Methods in Iteration", Proc. Amer. Math. Soc. 4 (1953), 506-510.

[18] Nashed, M. Z., "Generalized Inverses, Normal Solvability, and Iteration for Singular Operator Equations", in Nonlinear Functional Analysis and Applications, (ed. Rall, L. B.) Academic Press, New York, 1971.

[19] Outlaw, C., and Groetsch, C. W., "Averaging Iteration in a Banach Space", Bull. Amer. Math. Soc. 75 (1969), 430-432.

[20] Petryshyn, W. V., "On Generalized Inverses and Uniform Convergence of $(I-\beta K)^n$ with Applications to Iterative Methods", J. Math. Anal. Appl. 18 (1967), 417-439.

[21] Reich, S., "Fixed Points Via Toeplitz Iteration", Bull. Calcutta Math. Soc. (to appear).

[22] Reinermann, J., "Über Toeplitzsche Iterationsverfahren und einige Anwendungen in der Konstruktiven Fixpunktheorie", Studia Math. 32 (1969), 209-227.

[23] Rhoades, B. E., "Fixed Point Iterations Using Infinite Matrices", Trans. Amer. Math. Soc. (to appear).

[24]   Schaefer, H., "Über die Methode sukzessiver Approxi-
       mationen", J. Deutsch. Math. Verein. 59 (1957), 131-
       140.

[25]   Showalter, D., "Representation and Computation of the
       Pseudoinverse", Proc. Amer. Math. Soc. 18 (1967), 584-
       586.

[26]   Showalter, D., and Ben-Israel, A., "Representation and
       Computation of the Generalized Inverse of a Bounded
       Linear Operator Between Two Hilbert Spaces", Accad.
       Naz. Dei Lincei 48 (1970), 184-194.

[27]   Zeller, K., Theorie der Limitierungverfahren, Spring-
       er-Verlag, Berlin, 1958.

Solving Economic General Equilibrium
Models by the Sandwich Method

James MacKinnon

The Sandwich Method is a complementary pivoting algorithm for finding approximate fixed points of continuous mappings from the unit simplex into itself, and for solving related problems. It has been used chiefly to solve economic
general equilibrium models, so far purely theoretical ones
not intended to correspond to any real economy. This paper
deals with certain aspects of computational experience with
the Sandwich Method. Most of the results should also apply
to other complementary pivoting algorithms; whether they also
apply to other types of problems is not yet known.

A full description of the Sandwich Method would be out
of place here; see Kuhn and MacKinnon [in preparation]. The
algorithm is essentially that of Kuhn [1968]: Integer labels
are used and the search takes place on a regular subdivision
of the simplex, which must be properly labelled in the usual
sense of Sperner's Lemma (a vertex cannot get label i if
its i-th co-ordinate is zero). Every vertex in the subdivision can be described as a vector of integers which sum to an
integer, D , which describes the subdivision; the mesh is
proportional to $1/D$. The Sandwich Method differs from
Kuhn's 1968 algorithm in that an artificial dimension is
added to the problem in order to provide a start close to
where the answer is expected to be. This technique was first
used by Merrill [1971], and was independently rediscovered.

As an illustration of how the Sandwich Method works,
consider Figure 1. It shows a 1-simplex with D = 12. There
are completely labelled subsimplices at A, B and C . The
Sandwich Method operates by embedding the 1-simplex of Figure
1 in the 2-simplex (with D = 13) which is shown in Figure 2;
only the bottom three layers are shown, since the rest never
play any part in the algorithm. All vertices on the upper

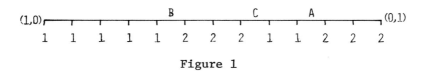

Figure 1

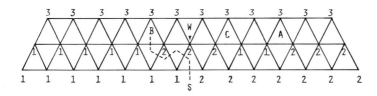

Figure 2

layer are given the label 3 (or, in the general case, n + 1 , where n - 1 is the dimension of the original simplex). Vertices on the lower layer are labelled so that there is a unique start at S . This is accomplished by choosing the vertex W and labelling vertices on the bottom layer by the rule

$$L(x) = k \quad \text{if} \quad (x_k - W_k) \geq (x_j - W_j) \quad \text{for all} \quad j \quad , \text{ and}$$

$$(1)$$

$$k < j \quad \text{if} \quad (x_k - W_k) = (x_j - W_j) \quad .$$

This rule creates a unique start around W . Since there is only one start, the algorithm must, by the usual Sperner's Lemma arguments, terminate at a completely labelled subsimplex. Because this must include a point on the upper layer, it must also include vertices with labels 1 to n on the middle layer. Thus the algorithm is guaranteed to find a completely labelled subsimplex of the original simplex, in this case B . So solving the artificial problem of Figure 2 also solves the real problem of Figure 1.

In the absence of outside information, the obvious, and also optimal, way to choose W is to pick an integer approximation to the barycentre of the simplex, which is just the point (1/n, 1/n, ..., 1/n) . It is possible, however, to acquire information at low cost by initially solving the problem for a small value of D . To achieve an accurate answer efficiently, the problem would then be solved succes-sively a number of times with increasingly large values of D , using the answer each time to choose W for the next. The obvious way to choose W is to make it an integer

approximation to the barycentre of the completely labelled
subsimplex found on the previous run. That choice has been
embodied in the algorithm.

In the remainder of the paper four aspects of computational experience are discussed. In the case of economic
general equilibrium models, many different labelling rules
can be used to solve any given problem. A fairly large-
scale experiment was performed to determine the relative
performance of five different rules, and computation cost
turned out to be quite sensitive to the choice. The problem
of how to increase D in an efficient manner is treated
theoretically, and the theory is tested against some experimental results. The relationship between dimensionality and
computation cost is also investigated experimentally. Finally, it is suggested, somewhat tentatively, that problems can
be divided into three different types according to the way
the Sandwich Method performs when solving them. A simple
modification is suggested which can substantially reduce computation time in one of these cases. This modification would
be applicable to any complementary pivoting algorithm utilizing an artificial layer to provide restarts.

Many economic general equilibrium models, though by
no means all, can be put into the form of what I shall call
the standard model. The standard model is completely described by n non-negative prices P_1 to P_n , which sum
to one and may be written as a vector, P , and by n excess
demand functions $g_i(P)$, which are continuous when all
prices are positive, bounded from below, and homogeneous of
degree zero in P . In addition, Walras' Law holds in the
strong form: for any vector of prices, the aggregate

value of excess demand , $\sum_{i=1}^{n} g_i(P)P_i$, equals zero. This

law, which may seem mysterious, follows very easily from the constraint that every agent must purchase goods to the same value as those he sells. Thus the value of each agent's excess demands must be zero, and summing over all agents yields Walras' Law. The definition of an equilibrium price vector in the standard model is a vector $\hat{P}$ such that $g_i(\hat{P}) \leq 0$ for all i and $g_i(\hat{P})\hat{P}_i = 0$ for all i . That is, there must not be positive excess demand for any good, and there may be excess supply only if its price is zero.

Consider the following rule for assigning integer labels to vertices in the price simplex:

$$L(P) = k \quad \text{if} \quad A_k(P)g_k(P) \leq A_j(P)g_j(P) \quad \text{for all } j \quad ,$$

$$(2)$$

$$P_j \quad \text{and} \quad P_k > 0, \quad \text{and} \quad k < j \quad \text{if} \quad A_k(P)g_k(P) = A_j(P)g_j(P)$$

where $A_i(P)$ is any non-negative function of prices, which is positive when P_i is positive. Rule (2) is clearly a proper labelling rule, because label i cannot occur when P_i is zero. Thus, by Sperner's Lemma, there must exist at least one completely labelled subsimplex in any simplicial subdivision of the price simplex. Each of these subsimplices has a barycentre, which is just the arithmetic mean of its vertices. If D is allowed to increase without limit, an infinite sequence of barycentres of completely labelled subsimplices will be generated. Because the simplex is a

compact set, there must exist a subsequence of this sequence
of barycentres which converges to a single point, P^* .
Since the mesh of the completely labelled subsimplices goes
to zero as D goes to infinity, P^* must be arbitrarily
close to points with all n labels, that is, to points
where $A_k(P)g_k(P) \leq A_j(P)g_j(P)$ for all k and for all j .
Since the excess demand functions are continuous, this implies that $A_k(P^*)g_k(P^*) \leq A_j(P^*)g_j(P^*)$ for all k and j ,
providing that $P_k^* > 0$ and $P_j^* > 0$. Obviously this can
only happen if the equality holds in all cases. But by
Walras' Law, that is only possible if $g_k(P^*) = g_j(P^*) = 0$.
So far, we have not dealt with the case where some elements
of P^* are zero. But it is clear that if $P_i^* = 0$,
$A_i(P^*)g_i(P^*) \leq 0$, for with some excess demands zero, P^*
could never get label i if the i-th excess demand were
positive. It is thus apparent that P^* must satisfy both
sets of equilibrium conditions: $g_i(P^*) \leq 0$ for all i ,
and $g_i(P^*)P_i^* = 0$ for all i .

The foregoing demonstrates that for an arbitrarily fine
subdivision, rule (2) generates a completely labelled subsimplex which is equivalent to an equilibrium. It should be
clear that, for a sufficiently fine but finite subdivision,
a completely labelled subsimplex will be near an equilibrium.
Thus the Sandwich Method may be used to calculate approximate
equilibria of general equilibrium models which are in the
form of the standard model. Rule (2) is actually a whole
class of labelling rules, depending on how the $A_i(P)$ are
chosen. The question which naturally arises when one goes

to compute equilibria is whether the choice matters.   The
next section will attempt to answer that question.

Alternate Labelling Rules

The internal operation of complementary pivoting algorithms such as the Sandwich Method has always seemed rather
mysterious to me.   I therefore find it very difficult to predict, from my knowledge of how the algorithm works, what
choices for $A_k(P)$, if any, are likely to work particularly
well.   However, it seems reasonable that there should be some
parallel between the performance of the Sandwich Method and
that of cruder algorithms for solving general equilibrium
models.   In the case of such cruder algorithms it will be
possible to say something about the expected performance of
various rules.   I hypothesize that if we associate these
rules from the crude algorithm with labelling rules, and test
the latter on a large number of problems, the empirical
rankings of the latter will be essentially the same as the
theoretical rankings of the former.

Let us consider, therefore, the crudest imaginable
algorithm for finding an equilibrium.   At the   i-th   step
there is a vector of prices, P^i , from which a vector of
excess demands, $g(P^i)$, is computed.   If these are not sufficiently close to zero, the algorithm may choose to lower
one and only one price, renormalizing the others to sum to
unity.   The basic problem facing the designer of the algorithm is which price to lower.   One possibility is to choose
the good in the greatest excess supply, a procedure which
will be referred to as rule 1.   The problem with rule 1 is
that some goods may tend to be in very large excess supply

if they are in excess supply at all, while other goods may
tend never to be in large excess supply. As an example of
the former, consider potato chips; as an example of the latter, consider jumbo jets. Surely an excess supply of ten
million potato chips is not more important than an excess
supply of ten DC-10's. But rule 1 would cause the price of
the former to be lowered. One alternative to rule 1 is to
pick the good with the greatest value of excess supply, a
procedure which will be referred to as rule 2. Rule 2 seems
much more reasonable than rule 1. But if the algorithm is
not near an equilibrium, prices may be misleading. Another
alternative, then, is to deflate excess supply by total supply, $s_j(P)$, before picking the good in greatest excess
supply. This will be referred to as rule 3. Again, if
prices are not near an equilibrium, total supply may differ
considerably from equilibrium supply. If P_j is too low,
one would expect $s_j(P)$ to be too low as well. Thus
$P_j/s_j(P)$ could be expected to vary less than P_j itself.
Another possible rule, then, is to both multiply by price
and divide by total supply, before picking the good in
greatest excess supply. This will be referred to as rule 4.

Rules 1 through 4 can easily be converted to labelling
rules: rule 1 is equivalent to setting $A_i(P) = 1$, rule 2
is equivalent to setting $A_i(P) = P_i$, rule 3 is equivalent
to setting $A_i(P) = 1/s_i(P)$, and rule 4 is equivalent to
setting $A_i(P) = P_i/s_i(P)$. Finally, we may add rule 5,
which involves setting $A_i(P) = s_i(P)/P_i$. Consideration
of the crude algorithm suggests that rule 5 should work

badly, so testing rule 5 should provide strong evidence for
or against my hypothesis that consideration of crude algorithms can provide a guide to the performance of the Sandwich
Method.   There are of course many other rules that could be
tried.   One variation would be to substitute total demand for
total supply in rules 3, 4, and 5.   I tried using $d_i(P)$ in
rule 3 in some preliminary experiments, and always got results identical or virtually identical to those with $s_i(P)$.
This is not surprising, because most of the time the algorithm is fairly close to an equilibrium, which means that
$s_i(P)$ is fairly close to $d_i(P)$. Thus rules using $d_i(P)$
have not been examined further.

In order to test rules 1 through 5, I constructed
three artificial equilibrium models with randomly chosen
parameters, which will be referred to as the exchange,
production, and taxation models in view of their main features.   The exchange model is identical in form to the models
solved by Scarf [1967].   The others are original.   All three
are described in detail in the Appendix.   For each of the
five labelling rules, I solved 100 randomly chosen exchange
models with n = 5 (n is the number of prices; n − 1 is
the dimensionality of the problem), 200 exchange models with
n = 4 , 100 production models with n = 5 , and 100 taxation models with n = 6 . The results are shown in Tables
1 and 2.

Table 1 is largely self-explanatory.   On the average,
rule 5 was in all cases notably worse than any of the others,
and rule 1 was notably worse than rules 2, 3, or 4.   The
average results presented in Table 1 conceal considerable
variation from problem to problem.   In all four experiments,

Table 1

Performance of Alternate Labelling Rules

	Rule 1	Rule 2	Rule 3	Rule 4	Rule 5
Exchange 1:					
sample size = 100					
n = 5					
D = 16 x 3^8					
Mean No. of labellings	124	107	107	104	156
Median No. of labellings	111	108	106	102	121
Exchange 2:					
sample size = 200					
n = 4					
D = 20 x 3^8					
Mean No. of labellings	72	69	69	67	82
Median No. of labellings	69	68	67	67	74
Production:					
sample size = 100					
n = 5					
D = 16 x 3^8					
Mean No. of labellings	278	228	240	215	320
Median No. of labellings	150	133	143	129	184
Taxation:					
sample size = 100					
n = 6					
D = 25 x 3^8					
Mean No. of labellings	277	221	201	205	***
Median No. of labellings	214	214	196	198	282

Notes:

Rule 1 was: $L(P) = k$ if $g_k(P) \leq g_i(P)$ for all i, P_i and $P_k > 0$.

Rule 2 was: $L(P) = k$ if $g_k(P)P_k \leq g_i(P)P_i$ for all i , P_i and $P_k > 0$.

Rule 3 was: $L(P) = k$ if $g_k(P)/s_k(P) \leq g_i(P)/s_i(P)$ for all i , P_i and $P_k > 0$.

Table 1 (continued)

Rule 4 was: $L(P) = k$ if $g_k(P)P_k/s_k(P) \leq g_i(P)P_k/s_i(P)$ for

all i, P_i and $P_k > 0$.

Rule 5 was: $L(P) = k$ if $g_k(P)s_k(P)/P_k \leq g_i(P)s_i(P)/P_i$ for

all i, P_i and $P_k > 0$.

***The sample mean in this case was several thousand, because four samples exhibited extreme behavior of the type referred to below as 'case 2', and took far more labellings to terminate than the other 96 samples together. This could have been avoided if the initial subdivision had been larger. The sample mean is not reported in the table, because in this case only the median is a reliable measure of central tendency.

Table 2

Pairwise Comparisons of Alternate Labelling Rules

	5 – 1	1 – 2	1 – 3	1 – 4	2 – 3	2 – 4	3 – 4
Exchange 1							
1. First wins	16	27	26	31	47	44	43
2. Tie	1	14	5	3	7	5	5
3. Second wins	83	59	69	66	46	51	52
(1/(1+3))	.16	.31	.27	.32	.51	.46	.45
z-ratio	6.74	3.45	4.41	3.55	-0.10	0.72	0.92
Exchange 2							
1. First wins	35	55	51	59	80	64	72
2. Tie	17	26	30	15	32	38	26
3. Second wins	148	119	119	126	88	98	102
(1/(1+3))	.19	.32	.30	.32	.48	.40	.41
z-ratio	8.35	4.85	5.22	4.93	0.62	2.67	2.27
Production							
1. First wins	4	10	26	16	76	42	18
2. Tie	1	1	6	0	2	2	2
3. Second wins	95	89	68	84	22	56	80
(1/(1+3))	.04	.10	.28	.16	.78	.43	.18
z-ratio	9.15	7.94	4.33	6.80	-5.45	1.41	6.26
Taxation							
1. First wins	8	35	23	30	27	32	56
2. Tie	1	1	2	0	2	1	3
3. Second wins	91	64	75	70	71	67	41
(1/(1+3))	.08	.35	.23	.30	.28	.32	.58
z-ratio	8.34	2.91	5.25	4.00	4.44	3.52	-1.52

there was for every rule, even rule 5, at least one problem
on which it outperformed all the others. This suggests that
comparing the performance of alternate labelling rules, or
different algorithms, on the basis of only a few test problems, can be seriously misleading.

Although some of the differences between the performance of various rules which are observed in Table 1 are
striking, it seems desirable to test statistically the hypothesis that these differences could have been due to pure
chance. Due to the extreme skew of some of the distributions
(note the differences between means and medians in some
cases), it seems likely that tests on means or other parametric tests which assume normality could be seriously misleading. I therefore used somewhat weaker but more robust
tests based on the binomial distribution. One measure of the
relative performance of two rules is the number of times each
requires fewer iterations than the other, when applied to the
same set of problems. The number of occasions on which the
first rule dominates, N_1 , should be distributed according
to the binomial distribution for $N_1 + N_2$ trials, with unknown p . The null hypothesis, that both rules perform
equally well, is that p = .5 . If the null hypothesis
holds, then the test statistic

$$z = \frac{N_1/(N_1 + N_2) - .5}{\sqrt{.25/(N_1 + N_2)}}$$

should have approximately the standard normal distribution
(and for samples as large as these the approximation should
be a good one). If the absolute value of z is greater than

1.96, we may reject the null hypothesis at the .95 level.

Table 2 shows the results of such pairwise tests for seven out of the ten possible pairs. In all experiments rule 1 dominates rule 5 in a statistically significant sense, and is in turn dominated by rules 2, 3 and 4. Results for the pairs 2-5, 3-5 and 4-5 can easily be deduced, and are therefore not included in the table. These results are quite striking, and strongly tend to confirm the predictions from consideration of the crude algorithm. From the latter one may also predict that rule 4 will dominate rule 2, because the $A_i(P)$ will vary less away from equilibrium, and indeed it does so, although the differences are significant in only two of the four cases. I was unable to make any predictions about the relative performances of 2 and 3, and 3 and 4. It turns out that they vary from experiment to experiment. Rules 2 and 3 perform equally well on the exchange problems, 2 dominates 3 on the production problems, and 3 dominates 2 on the taxation problems. Rule 4 dominates rule 3 significantly in two cases and insignificantly in one, but is in turn insignificantly dominated once.

It seems safe to draw the following conclusions. First, rules 1 and 5 work badly and should be avoided. Second, rule 4 seems to work particularly well. Third, consideration of cruder algorithms provides great insight into how alternative labelling rules will work. This procedure might well be applied to other types of problem as well.

Increasing D in an Optimal Fashion

In the experiments reported on above, a large final subdivision was always achieved by multiplying a small

initial D by three a number of times. This procedure was
not settled on at random. The following analysis of how to
choose an efficient sequence of subdivisions suggests that it
should be a good one. Although this analysis involves some
fairly drastic simplifying assumptions, it yields remarkably
concrete results which are not contradicted by the evidence.

Suppose that we wish to terminate at some prespecified
final subdivision, D_f , in a minimum number of labellings
(where 'labelling' does not refer to artificial labellings,
since they are very cheap). We must first choose some initial subdivision, D_o , and then increase it a number of
times. If the characteristics of the problem do not change
as the mesh gets finer, it should be optimal to increase D
by the same factor r each time; thus $D_i = rD_{i-1}$ always.
To make the problem tractable, I shall assume that this is
the case. By definition, then,

$$D_f = D_o r^N \tag{3}$$

where N is the number of times that D is increased. (3)
can be rewritten as

$$N = \log (D_f/D_o)/\log r \quad . \tag{4}$$

In practice, D_o and N must be integers, and it will simplify the programming if r is constrained to be an integer
as well; that is how the Sandwich Method is currently programmed. I shall, however, ignore the integer constraints
here. Provided that D_f is large enough, only the constraint

on r should be important, since rounding D_o and N to

the next higher integer will merely cause D_f to be overshot

slightly.

On each incrementation the algorithm starts near the
barycentre of a completely labelled subsimplex for subdivision D_i and ends at a completely labelled subsimplex for

subdivision rD_i . It seems reasonable that, for any given

problem, the cost of doing so should depend only on r .
For concreteness, I have assumed that the number of labellings required on each incrementation is

$$c \, r^\alpha \, e^u \qquad\qquad (5)$$

where u is an error term. To test this assumption, (5) was
estimated in log-linear form by ordinary least squares, using
some two hundred observations from each of six text problems.
In all cases the fit was impressive $(R^2 = .5)$, and the
residuals did not suggest misspecification. My prior expec-
tations were that α should be unity. In fact, the estimates varied between .5 and 1.1, and were significantly less
than unity in half the cases. Clearly, then, α can be expected to vary from problem to problem. It also seems reasonable to assume that the cost of the initial solution varies linearly with D_o . I have not tested this assumption,

but Scarf [1973] presents evidence to show that it works well
for his algorithms.

On the assumptions of the previous paragraph, the expected total cost of solving a given problem for final subdivision D_f is

$$Ncr^{\alpha} + dD_o \qquad (6)$$

where c has been redefined to equal $cE(e^u)$. Using (4),
(6) becomes

$$\log \ (D_f/D_o)c(r^{\alpha}/\log r) + dD_o \ . \qquad (7)$$

Minimizing (7) with respect to r is clearly equivalent to
minimizing

$$r^{\alpha}/\log r \ . \qquad (8)$$

Differentiating (8) with respect to r and equating the derivative to zero yields

$$\frac{\alpha r^{\alpha-1} \log r - r^{\alpha-1}}{(\log r)^2} = 0 \qquad (9)$$

from which it is easily shown that

$$\log r = 1/\alpha \ . \qquad (10)$$

This is a remarkably simple result. There is a unique optimal rate of incrementation, and it is generally quite small:
if $\alpha = 1$, $r = e$; if $\alpha = .8$, $r = 3.49$; if $\alpha = .6$, $r = 5.31$;
and so on. It is not generally as low as two, however; for
that, α would have to be approximately 1.4, which is much
higher than any of my estimates. If r is constrained to be
an integer, we can easily maximize it by evaluating (8) for

reasonable values. It turns out that, for α between .80
and 1.14 , three is the optimal value for r .

We can maximize (7) with respect to D_o in a similar
fashion. The result is

$$D_o = (c/d)\ r^{\alpha}/\log r\ .\qquad(11)$$

In effect, then, D_o just varies linearly with c and in-
versely with d . The more expensive incrementation is, the
larger D_o should be; the more expensive the initial search
is, the smaller it should be. Thus if there is prior information about where the solution is expected to be, so that
the initial search will be cheap, D_o should be made relatively large. My experience suggests that in general, however, D_o should be quite small - twenty-five or less in
most cases.

These results are interesting. They suggest that refining the mesh by a factor of two each time, which is an essential part of current algorithms that refine the mesh homotopically, such as Eaves and Saigal [1972], is generally not
optimal. But they also suggest that the mesh should not be
refined too quickly. The question which naturally arises at
this point is whether these results are vitiated by the unrealistic assumptions used to derive them. I have gathered
some experimental evidence which suggests that they are not.

The following sequences of subdivisions yield final
subdivisions which are of almost equal size: 18×2^{16} ,
20×3^{10} , 18×4^{8} , 15×5^{7} , 25×6^{6} , and 20×9^{5} .

I solved forty exchange models using labelling rule 4 and
each of these sequences, twenty problems with n = 4, and
twenty with n = 8. The results are shown in Table 3, and
they accord remarkably well with the theory; both r = 2
and r = 9 work badly compared with the other four rates,
and r = 5 does best on both sets of problems. This would be
consistent with an α of between .60 and .65 . For
both experiments, if mean or median numbers of labellings were
graphed against r , the graph would have the same sort of
bowl shape as expression (8). Thus the experimental results
seem to support the theory.

Dimensionality Versus Cost of Solution

There has been a good deal of speculation in the folk-
lore of complementary pivoting algorithms that the cost of
solution goes up as the square of the number of dimensions.
The random models created to investigate alternate labelling
rules could easily be varied in size, so I created data to
test this hypothesis by solving a number of similar models of
various sizes. The dependent variable in all cases was the
number of observations needed to increase D by a factor of
three; each random problem was used to generate eight observations. For the exchange and taxation models 280 observations were generated, 56 for each of n = 3, n = 6, n = 9,
n = 12 and n = 15 . The same was done for the linear
problem investigated by Scarf [1973]: find x non-negative
and summing to unity such that x = Cx , where C is a
square matrix with columns sums of unity. For the production
models, which were more expensive to solve, n = 15 was omitted to conserve funds. The equations which were fitted had

Table 3

Comparing Different Rates of Incrementation

RATE (D_i/D_{i-1}):	2	3	4	5	6	9
Initial subdivision:	18	20	18	15	25	20
Number of increments:	16	10	8	7	6	5
Final subdivision:	1179648	1180980	1179648	1171875	1166400	1180980

Exchange 3

$n = 4$

sample size = 20

	2	3	4	5	6	9
mean number of labellings:	111	80	79	76	85	92
median number of labellings:	111	80	79	77	80	88
number of times rate was fastest:	0	3	4	10	3	0

Exchange 4

$n = 8$

sample size = 20

	2	3	4	5	6	9
mean number of labellings:	475	421	411	398	425	549
median number of labellings:	475	445	404	400	419	556
number of times rate was fastest:	1	3	6	7	3	0

the form:

$$\log \text{LAB} = a_1 + a_2 \log(n-1) \quad . \qquad (12)$$

Results are shown in Table 4.

Table 4

Cost Versus Dimensionality

	a_1	a_2	R^2	s.e.e.
1. Linear models (x=Cx)	-.0857	1.9166	.9017	.4368
	(.0761)	(.0379)		
2. Exchange models	-.1022	1.9450	.9071	.4300
	(.0749)	(.0374)		
3. Taxation models	-.2538	2.1499	.8981	.5012
	(.0868)	(.0428)		
4. Production models	.2045	1.8209	.8674	.4598
	(.0866)	(.0478)		

Note: Figures in brackets are standard errors.

The maintained hypothesis is that $a_2 = 2$. In all four cases a_2 was reasonably close to two, but in three out of four it was significantly different therefrom. Thus the relationship between dimensionality and cost apparently varies from problem to problem, although not by a great deal for these examples. It is interesting to note how similar the equations are for the linear and exchange models; this suggests that the latter are more or less linear. This is consistent with other evidence that the exchange models are unusually well-behaved.

Table 4 should not be regarded as the final word on the
subject. The residuals would not have passed a test for
normality; they were skewed to the right, and tended to be
unusually small when n = 3 . Substituting n , which is
theoretically not the appropriate variable, for n - 1 ,
which is, tended to improve the R^2 slightly in all cases,
a phenomenon which I find very hard to explain. On those
equations, of course, a_2 was always well above two. Thus
it may be that the relationship between dimensionality and
cost is more complicated than the results reported in Table 4
suggest. It is, however, clear that cost rises rapidly with
dimensionality, and the $(n-1)^2$ rule does not seem to be a
bad approximation.

Computational Characteristics of General Equilibrium Problems

My computational experience, which is now rather extensive, suggests that economic general equilibrium models tend
to fall into three different classes, according to the way
the Sandwich Method behaves when solving them. For the case
where n = 3 , it is possible to graph the labelling associated with a problem, and I have done so in a number of cases.
There appears to be a close association between the way the
graph looks and the way the algorithm behaves, and I conjecture that these graphs also give some insight into the way
the algorithm performs on higher dimensional problems.

The most common type of computational behavior is what
I refer to as case 1. Figure 3 shows a typical graph for a
case 1 problem with n = 3 . The small triangle is the only

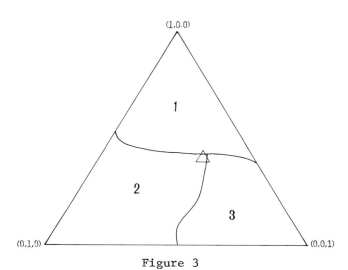

Figure 3

Computation: Case 1

completely labelled triangle for D = 16 . Note that it is
very close to the equilibrium (which of course is the point
where all three labelled regions come together). This is
bound to happen, because in the neighborhood of the equilibrium all regions are thick, and if the triangle were moved
much in any direction it would be unable to span one region
into the other two. Thus case 1 problems tend to be very
well-behaved, with the completely labelled subsimplex for subdivision D_i always lying close to the completely labelled
subsimplex for subdivision D_{i-1} . The exchange models always exhibited this type of behaviour, for example.

Case 2 behavior seems to occur much less frequently
than case 1. Figure 4 shows what sort of graph might give

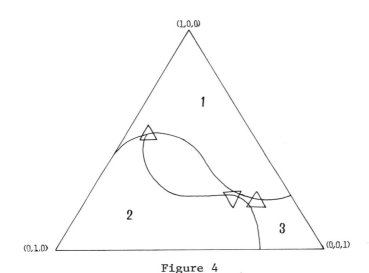

Figure 4

Computation: Case 2

rise to this type of behavior.   The problem is that for low
subdivisions completely labelled subsimplices which are not
near equilibria may exist, but than vanish when the subdivision becomes sufficiently fine.   The algorithm may then have
to move a long way to find another completely labelled subsimplex.   Algorithms which refine the mesh by using an artificial layer, such as the Sandwich Method, will inevitably
have trouble with case 2 problems.   One would expect algorithms which refine the grid homotopically, such as that of
Eaves and Saigal [1972], to perform better on such problems.

  Case 3 behaviour, in my experience, is much more common than case 2, although it does not appear to any extent
among the very simple test problems reported on in this paper.
The essence of case 3 behaviour is that the completely

labelled subsimplex tends to move a lot, usually in a systematic way, every time D is increased. Figure 5 shows
what sort of graph might give rise to this type of behaviour.

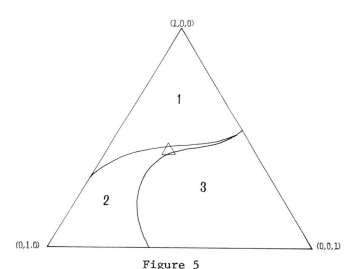

Figure 5
Computation: Case 3

The essential thing is that, in the neighborhood of an
equilibrium, at least one region of labels should be very
thin, so that the probability of getting a completely labelled subsimplex near the equilibrium when the mesh is coarse
is very low. As the mesh get finer, the completely labelled
subsimplex tends to slide along the narrow region towards
the equilibrium, moving less and less each time in real
terms, but often moving roughly the same distance in terms
of the integer grid. This sort of behaviour can increase

computation cost dramatically.

If the completely labelled subsimplices tend to move systematically, it is obviously not optimal to choose the barycentre of the completely labelled subsimplex for D_{i-1} as the start for D_i . I have therefore modified the algorithm so that W^k , the starting point on the k-th incrementation of D , may be moved somewhat according to the relationship between the completely labelled subsimplices on the two previous incrementations. This procedure is rather ad hoc, but seems to work well. Define

$$\bar{x}^k = \sum_{i=1}^{n} x_i^k \tag{13}$$

where x_i^k is the i-th vertex of the completely labelled subsimplex on the k-th incrementation. Now define

$$D_1 = \bar{x}^{k-1} - r\,\bar{x}^{k-2}$$

$$\text{and} \tag{14}$$

$$D_2 = \bar{x}^{k-2} - r\,\bar{x}^{k-3}$$

where r , as before, is the rate of incrementation. W^k is then chosen as an integer approximation to

$$(K_o r \bar{x}^{-k} + K_1 D_1 + K_2 D_2)/K_o n \tag{15}$$

where K_o , K_1 and K_2 are non-negative weights chosen by the user. If K_1 and K_2 are zero, W^k is just the usual barycentric restart. The larger K_1 and K_2 are relative to K_0, the more W^k will be influenced by extrapolation from previous moves. There is no reason to believe that K_1 and K_2 should sum to K_0. If case 3 is very pronounced, they should perhaps sum to more than K_0, while if it is not pronounced they should certainly sum to less than K_0. Much of the movement on previous incrementations is purely random, and making K_1 and K_2 sum to less than K_0 is a way of discounting it. Of course, if a problem displays pure case 1 behaviour, all such movement is random, and making K_1 and K_2 non-zero will simply increase computation time, because they will always tend to pull W^k away from the optimal place to start, which is at the barycentre of the previous completely labelled subsimplex.

Table 5 provides some illustration of how effective this procudure can be. The problem in question was considerably more complicated than any of the random parameter problems discussed in the Appendix, and exhibited pronounced case 3 behaviour. Using the conventional restart, K_1 and K_2 set to zero, this problem required 646 labellings to solve; using extrapolative restarts, it required no more than 467, and as few as 215. These are substantial savings.

This procedure for dealing with case 3 problems is of course only one of many that might be adopted. I do not contend that it is the best one. The point is simply that complementary pivoting algorithms which use an artificial layer

Table 5

Extrapolative Restarts

Weights:	K_0	K_1	K_2	Number of Labellings
	1	0	0	646
	1	1	0	361
	2	1	0	445
	2	1	1	467
	3	1	1	448
	3	2	1	235
	4	2	1	361
	5	2	1	362
	5	3	2	456
	5	4	1	395
	5	4	2	336
	10	7	3	215

$n = 5$

$D = 16 \times 3^9$

for restarts can be made a good deal more effective at
dealing with case 3 problems, and may therefore have an advantage over algorithms which refine the grid homotopically.
How important this is likely to be will depend on how frequently such problems occur in practice.

If my experience is not unusual, there would seem to be
a place both for algorithms that refine the grid homotopically and for algorithms that use an artificial layer to provide
restarts. The former can be expected to work better on case
2 problems, but the latter, when extrapolative restarts are

used, should work better on case 3 problems. Both types of
algorithms can be expected to work well on case 1 problems.
Future comparisons of the effectiveness of different algorithms should take the characteristics of the test problems
used into account.

REFERENCES

[1]   Eaves, B. Curtis "Homotopies for the Computation of
      Fixed Points," Mathematical Programming, 3, 1972, 1-22.

[2]   Eaves, B. Curtis, and Saigal, R.   "Homotopies for the
      Computation of Fixed Points on Unbounded Regions,"
      Mathematical Programming, 3, 1972, 225-237.

[3]   Kuhn, H. W.   "Simplicial Approximation of Fixed Points,"
      Proceedings of the National Academy of Sciences, 61,
      1968, 1238-1242.

[4]   Kuhn, H. W., and MacKinnon, James   "The Sandwich Method
      for Finding Fixed Points," in preparation.

[5]   MacKinnon, James   "Urban General Equilibrium Models and
      Simplicial Search Algorithms," Journal of Urban Econom-
      ics, 1, 1974, 161-183.

[6]   Merrill, Orin H.   "Applications and Extensions of an
      Algorithm that Computes Fixed Points of Certain Non-
      Empty Convex Upper Semi-continuous Point to Set Map-
      pings," Department of Industrial Engineering, University
      of Michigan, Technical Report No. 71-7, 1971.

[7]   Scarf, H. E.   "The Approximation of Fixed Points of a
      Continuous Mapping," SIAM Journal on Applied Mathemat-
      ics, 15, 1967, 1328-1343.

[8]   Scarf, H. E., with the collaboration of Hansen, T.   The
      Computation of Economic Equilibria, Yale University
      Press, New Haven, 1973.

[9]   Shoven, John B., and Whalley, John   "A General Equilib-
      rium Calculation of the Effects of Differential Taxation
      of Income from Capital in the U.S.," Journal of Public
      Economics, 1, 1972, 281-321.

[10] Wilmuth, Richard J. "The Computation of Fixed Points,"
      Doctoral Dissertation, Department of Operations Re-
      search, Stanford University, 1973.

APPENDIX

This appendix provides brief technical descriptions of the three classes of random parameter equilibrium models which were used in the experiments described in the test.

1. Exchange Models

There are n goods and m individuals. In the experiments reported on in Table 3-3, m was two; elsewhere it was three. The demand for good i by individual j is

$$D_{ij} = \frac{A_{ij}\left(\sum_{k=1}^{n} W_{kj}P_k\right)}{P_i^{B_j}\left(\sum_{k=1}^{n} A_{kj}P_k\right)^{1-B_j}}$$

and the supply of good i by individual j is W_{ij} . The excess demand for good i is

$$g_i = \sum_{j=1}^{m} (D_{ij} - W_{ij}) \ .$$

Random parameters were chosen as follows: A_{ij} was uniform $(0.5 - 5.5)$; B_j was uniform $(0.4 - 2.0)$; and W_{ij} was uniform $(0.0 - 20.0)$.

2. Production Models

There is one individual, n-1 produced goods, and one factor (good n). The final demand for good i is

$$D_i^f = (10^6 P_n + \sum_{k=1}^{n-1} \Pi_k) B_i / P_i \quad .$$

The production function for good i is

$$S_i = R_i \left[F^{A_{in}} \prod_{k=1}^{n-1} x_k^{A_{ik}} \right]^{.97}$$

where F is the input of the factor, and x_k is the input of the k-th produced good. Supplies of good i , demands for F and for x_k , and profits Π_i are all determined by maximizing

$$\Pi_i = P_i S_i - P_n F - \sum_{k=1}^{n-1} P_k x_k \quad .$$

This is easily done, but the equations which result are messy and will therefore not be reported here. Random parameters were chosen as follows: R_i was uniform (5.0 - 15.0); B_i was uniform (0.1 - 1.1); for k ≠ n, A_{ik} was uniform (0.001 - 1.001) divided by $\sum_{k=1}^{n} A_{ik}$; A_{in} was uniform

$((n-1)/10 - (11/10)(n-1))$, divided by $\sum\limits_{k=1}^{n} A_{ik}$. The way the A_{ik}'s were chosen may seem odd. This procedure was followed so that demand for the factor did not tend to go to zero as n increased.

3. Taxation Models

There are 2 individuals, 2 goods, and n-2 taxing authorities, each of which issues a different type of tax ticket. P_1 and P_2 are the prices of the goods, while P_i for $i \geq 3$ is the price of a tax ticket. The demand for good i by individual j is

$$D_{ij} = \left(\sum_{k=1}^{n} W_{kj} P_k \right) \bigg/ \left[P_{\bar{i}j}^* \left(\frac{P_{ij}^* A_{\bar{i}j}}{P_{\bar{i}j}^* A_{ij}} \right)^{S_j} + P_{ij}^* \right]$$

where $P_{ij}^* = P_i + \sum\limits_{k=1}^{n-2} T_{ijk} P_i$ and $\bar{i} = 2$ if $i = 1$, and $\bar{i} = 1$ if $i = 2$. The demand for tax ticket k by individual j is

$$D_{kj} = \left(\sum_{i=1}^{2} D_{ij} T_{ijk} P_i \right) / P_k$$

and the supply of good or ticket i by individual j is W_{ij} . Random parameters were chosen as follows: T_{ijk} was uniform (0.0 - 0.5); S_j was uniform (.1 - 3.1); A_{ij} was uniform (.1 - 1.1); and W_{ij} was uniform (100,000.0 - 1,000,000.0) .

Applying Fixed Point Algorithms to the
Analysis of Tax Policies

John B. Shoven

1. Introduction

This paper has three related purposes. First, it offers a nontechnical summary of the techniques which John
Whalley of the London School of Economics and I have developed for incorporating taxes and tariffs into a general equilibrium computational methodology. Our work, which has been
based on Scarf's algorithm [7], is applicable to alternative
fixed point algorithms. Second, this paper reviews the original economic application we jointly made of this technique--
evaluating the impact of the distortionary taxation of capital income in the U.S. and presents a portion of some later
results of mine on that topic. Lastly, the possibilities for
further applications are briefly discussed and promising research avenues surveyed.

The chief advantage of the algorithmic approach over
others in economic policy evaluation is that it does not require any localization or linearization assumptions and
therefore offers an appropriate tool for analyzing large non-
marginal changes. This feature is not shared by most alternative models in economics which are based on calculus. In
these analytic models, parameters such as demand elasticities
are only evaluated locally in the current observable economic
situation. The models typically make many linear or local
assumptions and are either not completely general in allowing
interactions between markets or they deal with "binary economies" (i.e., 2 goods, 2 factors, 2 consumers, 2 ...). The
algorithmic approach not only permits the analysis of large

policy changes, but also is not restricted to such binary economies. This approach has already proven useful in several applications, and its potential in evaluating large policy alterations seems very great.

2.   The Inclusion of Taxes into Scarf's Algorithm

This paper will not review the fundamentals of Scarf's algorithm or other fixed point algorithms, but will concentrate on the modifications required in order to use these techniques for tax policy evaluation. The notation and approach is consistent with that in Scarf [7], which contains a thorough and clear description of his algorithm. Basically, Scarf's algorithm and the other fixed point algorithms, when applied to general equilibrium economics, find a vector of commodity prices which constitutes an approximate equilibrium vector in that the supply of every commodity is approximately equal to its demand and firms are maximizing profits at these prices in determining their production. Nonincreasing returns to scale in production is required. In fact, production technology is commonly described using a matrix A

$$
A = \begin{bmatrix}
-1 & 0 & \cdots & 0 & a_{1,N+1} & \cdots & a_{1,M} \\
0 & -1 & \cdots & 0 & a_{2,N+1} & \cdots & a_{2,M} \\
\cdot & \cdot & & \cdot & \cdot & & \cdot \\
\cdot & \cdot & & \cdot & \cdot & & \cdot \\
\cdot & \cdot & & \cdot & \cdot & & \cdot \\
0 & 0 & \cdots & -1 & a_{N,N+1} & \cdots & a_{N,M}
\end{bmatrix}
$$

where each column represents a feasible activity and where
outputs have positive coefficients and inputs negative coefficients. As shown above there are a total of N commodities
in the model, and the first N columns of A simply indicate
the feasibility of free disposal of each of these commodities.
The only restriction placed on A is that it be such that the
set of x satisfying $Ax + W \geq 0$ be bounded, where W is
the N-dimensional column vector of the economy's initial endowments. This excludes the possibility of input free production and may be interpreted as guaranteeing that the production possibilities frontier is finite in all N dimensions.

        This description of production technology simplifies
some mathematical arguments and clearly is structurally based
on an input-output type of analysis. However, when evaluating
the impact of alternative economic policies, one often de-
sires to estimate small changes in the input ratios of the
various output sectors (i.e., so-called "factor-substitution"
effects). This would seem to require either a very long list
of feasible activities or an alteration in the algorithm to
allow the incorporation of continuous production functions.

        Since both consumer and producer reactions in general
equilibrium models are homogeneous of degree zero with respect
to prices, only relative prices are important and therefore
prices can be and are normalized to sum to unity. In order
to locate an approximate equilibrium price vector, a fine
grid of price vectors (or "candidates") is set up on the unit
N-simplex, and these algorithms can be thought of as a systematic means of moving through the grid vectors until an
answer is found. The rules of movement require that some information about each price vector in the grid be available if

and when that vector is considered. This information, which tells something about the economy at those prices, is contained in an N-dimensional "corresponding vector". If any of the prices are zero, then the corresponding vector is the negative of a disposal activity involving one of the free goods. If none of the prices are zero, then the profitability of each activity vector is calculated and if the most profitable activity actually earns a positive profit, then the corresponding column vector is the negative of that most profitable activity. If, on the other hand, even the most profitable activity in the A matrix loses money, then the corresponding column vector is the vector of total market demands at those prices. Given this information, the algorithm searches through the grid in a manner described in Scarf [7] and (empirically) quickly finds an approximate equilibrium price vector.

As described above, given any price vector, one must be able to determine the most profitable production activity. This does not require an activity analysis description of technology, however. Continuous production functions such as those with a constant elasticity of substitution (CES) can easily be incorporated. With production functions such as these, the optimal (i.e., cost minimizing) inputs for, say, one unit of output can be derived as an analytic function of input prices. For example, if the production function for a particular output is given as

$$Q = \gamma[\alpha L^{-\rho} + (1-\alpha) K^{-\rho}]^{-1/\rho} \tag{1}$$

and the input prices are P_L and P_k , the optimal inputs for one unit of output are

$$L = \gamma^{-1} \left[(1-\alpha) \left\{ \frac{\alpha P_k}{(1-\alpha)P_L} \right\}^{\rho/\rho+1} + \alpha \right]^{1/\rho} \tag{2}$$

and

$$K = \gamma^{-1} \left[\alpha \left\{ \frac{(1-\alpha)P_L}{\alpha P_k} \right\}^{\rho/\rho+1} + (1-\alpha) \right]^{1/\rho} \tag{3}$$

Given a vector of all prices, then, an optimal unit activity can be computed for each output or sector. Only these optimal activities need be considered in determining the corresponding vector to a price or "candidate" vector. If the most profitable of the optimal activities has a positive profit, then the negative of this activity becomes the corresponding vector. Otherwise, the corresponding column is the column of market demands evaluated at that price vector. If the price vector has a zero component, the same disposal activity as before is the corresponding vector. This technique obviously allows for very small input alterations in response to policy changes and has proven to be extremely useful in Whalley [15] and Shoven [8].

The next modification is the inclusion of taxes themselves. In particular, Whalley and I wanted to introduce discriminatory factor taxes -- i.e., to face different producers with different tax rates on their various inputs. In fact, as will be made clear, we can now handle a much broader range of taxes. One might at first think that including differential factor taxes simply involves having the jth producer pay $P_\ell \left(1 + t_\ell^j \right)$ for the use of the ℓth input, where t_ℓ^j is the tax rate the jth producer faces on commodity ℓ .

The problem with this involves incorporating the resulting government tax revenue. Given a set of all N prices, how is the government tax revenue to be determined? The tax revenue clearly depends upon the decisions of the various agents (producers in this case), but presumably the agents' decisions depend upon the government revenue and what is done with it. In general, then, commodity prices alone no longer convey sufficient information for agent decision making.

In order to simplify, assume that the total government revenue, R , is distributed to consumers. In fact, let the ith consumer have claim to a fraction, α_i , of the revenue where the α_i's sum to unity. With these assumptions, one method of including a discriminatory tax on an input, say, capital, is to create a fictitious commodity which can be termed "capital tax tickets". The firms in each taxed sector would face given ticket requirements per unit of capital employed. Each consumer, say the ith , would be endowed with an α_i fraction of the total number of tickets and these tickets would then be sold in an organized market. An individual consumer's share of the ticket proceeds (i.e., the tax revenue) would equal his share of the tickets. Since these artificial tickets would not directly enter into any consumer's utility function, they all would be sold. If one wanted to allow the government to retain some revenue for purchases of publicly provided goods and services, it would be treated as a consumer and endowed with some fraction of the total number of tickets. So, with this approach (which was used in Shoven and Whalley [11], a tax on a single commodity can be incorporated by creating a single additional (although fictitious) commodity.

There are several shortcomings of the above approach. First, the effective tax rate implied by a particular ticket requirement depends upon the price of the tickets. This, of course, is not something which can be imposed, but is a result of the algorithm. So, in order to establish particular capital tax rates, one is forced to iterate on the total number of tickets endowed until the price of a ticket is appropriate and establishes the desired tax rates. Second, in order to tax several different commodities, several different types of tickets must be created. This causes severe difficulties in imposing the rates, in that the number of commodities (or dimensionality of the problem) grows thus causing increasing computation costs.

A far superior technique for including taxes is thoroughly described in Shoven and Whalley [12]. The problems mentioned above are resolved by this method. Now, not only are all prices taken as given by consumers and producers in making their respective response decisions, but likewise the total government revenue is taken as given information. That is, each agent takes $(P_1, \ldots, P_N, R)$ as given and acts accordingly. With this method, each producer and consumer may face a different tax rate on each different commodity. The taxes, however, must be proportional to commodity usage, output, or consumption. In equilibrium, not only must the demand and supply be equated for each commodity, but R, the government revenue upon which decisions have been based, must in fact be collected. In order for the algorithm to achieve this additional equality, the corresponding vectors are augmented by an additional term. As one might expect, the addition involves the amount of tax payments collected with the particular price vector. If the column is the negative of a

production activity, then this vector is augmented by the
amount of taxes paid per unit level of operation of this
activity at those prices and given the tax rates applicable
on that activity. If the corresponding vector is the vector
of market demands at those prices, then the added coefficient
is the total amount of consumer taxes owed given $(P_1, \ldots, P_N, R)$
minus the announced government revenue R . With this additional information, it can be shown that the algorithm can always find an equilibrium. Again, the dimensionality is increased by one, but with this improved method all commodities
can be taxed at different rates for different agents. In addition, the tax rates can be directly imposed. This technique
has been used extensively in Whalley [15] and Shoven [8].

  With taxes added in this latter manner, extending the
approach to several countries is feasible and is discussed in
Shoven and Whalley [13]. Agents take as given all prices and
all government revenues and equilibrium requires that each
government collect its announced revenue. The corresponding
columns have an additional term for each country as might be
expected. It is required that production involve only commodities located in a single country, but otherwise no additional restrictive assumptions are necessary. The proof that
the method always will find an approximate equilibrium is
somewhat complicated and is presented in our article. Tariffs may be imposed on foreign goods purchased and consumers
may be endowed with both foreign and/or domestic assets.
This type of model seems most appropriate for evaluating the
impact of tariff reforms or the effects of the formation or
expansion of a customs union.

  While not previously discussed, import quotas, rather
than tariffs, can also be easily added to the model. The

technique would be similar to our first attempt at incorporating taxes in that an artificial (or, perhaps, sometimes
real) commodity, import rights or coupons, would be created.
The number of these rights would equal the import quota limit
and importers would be required to use one right for each
unit sold in the country imposing the quota. The rights
could either be auctioned off by the government or given to
importers, depending on what policy the government follows.
Either could be incorporated into this model.

Another modification which Whalley and I have made in
the computational algorithm recognizes the fact that when
different tax systems are being compared one often wants to
be certain that each system generated the same real yield.
In Shoven and Whalley [14] a procedure is described whereby
the algorithm not only determines equilibrium prices, but
also determines the rate structure required to match a particular revenue, say the revenue generated by some other system.
The technique does require that only one rate be endogenous
and that the others be taken as given. Clearly there is a
degree of arbitrariness as to which is taken as the rate to
be determined. The endogenous rate may, of course, be a
base rate upon which the other taxes (or surtaxes) are
"floated". In contrast with our other procedures, it is not
guaranteed that a solution exists in this framework, although
this has not diminished the usefulness of the technique for
empirical research. To cite an extreme example of nonexis-
tence, clearly it is impossible to generate an equal yield
with a tax on, say, rubber bands as with the entire U.S. corporation income tax. Nonetheless, this technique has proven
useful in comparing relevant alternatives and is used in both
Whalley [15] and Shoven [8].

As is well known, the computation time and, hence, costs of these algorithms rise rapidly with both the fineness of the grid of "candidate" equilibria and with the number of commodities. It has been observed that while many of the models dealt with for policy analysis are not globally stable for a gradient method or Newton's method technique (and hence the need for the algorithmic approach), they are locally stable using these methods. In Shoven [8], two alternative local termination routines similar in nature to Newton's method are presented, one based on Robinson [5]. These work extremely rapidly and allow one to work with very coarse grids, simply using the algorithm to determine a reasonable initial point for these termination routines. Of course, if the proposed change is sufficiently small, the current (i.e., before-change) observed equilibrium can be used as the starting point and the algorithm may not be necessary at all. The use of these linear termination routines in conjunction with coarse grids greatly diminishes the computation costs required for an accurate answer.

Another method of lowering computation time is possible under some conditions. With single-output continuous production functions for every output and with no intermediate inputs, output prices can be analytically derived from factor prices as long as each output is produced at a positive level. This makes it possible to work only with factor markets, deriving all other prices from factor prices and imposing equilibrium in output markets. Both the linear termination routines and this technique of reducing the fundamental dimensionality to that of the number of factors were used in Whalley [15] and Shoven [8].

The remaining technique related developments which Whalley and I have worked on involve parameterizing such general equilibrium models. One criticism of them is that it is difficult to estimate the number of parameters involved. The appropriate reply seems to be that the underlying economy modeled with a partial equilibrium approach involves just as many parameters -- it is just that most of them are arbitrarily taken to be zero. Whalley and I have observed that if one constrains the model to reproduce the observed economy when faced with the observed (tax) environment, the number of free or independent parameters is substantially reduced, often by a factor of about three. The remaining parameters must be extraneously derived. Heuristically, then, what we propose and have done is, given some extraneous parameter estimates and the observed economic variables, solve the model "backwards" to determine the necessary values of the remaining parameters (i.e., what they must equal to give such results). Then, with all of the parameter values known, a policy change can be considered and the model can be solved "forwards" and predict the new equilibrium. The biases and other econometric properties of such an approach have yet to be established, but when faced with a policy problem, it seems to offer a reasonable way of proceeding.

3. The U.S. Taxation of Capital Income

As an initial application of this methodology, John Whalley and I choose to evaluate the efficiency and incidence of the U.S. taxation of capital income. Our study appears in Shoven and Whalley [11]. Since that publication, the data on which it was based (which was the same as that used in Harberger [4]) have been found incorrect and have been revised

and the algorithmic techniques have been improved (this paper used the artificial and somewhat clumsy "capital tax
tickets" previously discussed). In Shoven [8] the problem is
looked at in greater depth with additional production disaggregation, improved data, and the more satisfactory technique
of incorporating taxes described in the previous section.
The new results, presented more completely in Shoven [10],
will be summarized here.

The importance of capital income taxes is undeniable.
Capital income amounts to roughly one fourth of all income
in the U.S. (even a larger fraction for most other countries),
and taxes on that capital income account for more than forty
percent of all federal, state, and local tax revenues. The
most important U.S. capital income taxes are the corporation
income tax, the local property tax, and state and federal income taxes as they apply to dividends and capital gains.
Despite the magnitude of these taxes, their impact is not
completely understood by public finance economists. In particular, the effects of the corporation income tax have been
subject to wide debate and many relatively recent economic
studies.

It is widely agreed that the capital income levies in
the U.S. are inefficient and distortionary because of the
fact that income from capital is taxed at different rates
depending on the location of the capital and depending on
the financial structure of the organization in which it is
a factor of production. In particular, capital income generated in the corporate sector is subject to the relatively
high 52% corporation income tax, while capital income from
the unincorporated sector is not. On top of this, property
tax rates vary by location and assessment practices, and

many industries (such as oil) are given special treatment
(e.g., depletion allowances). The fact that capital income
is taxed differently across production sectors tends, as one
would expect, to result in a misallocation of capital at some
cost to society. Several economists, Arnold Harberger [2, 3,
4], most prominent among them, have tried both to measure
this social cost or waste and to determine who bears the burden of this cost.

In his articles, Harberger divides the production side
of the economy into two sectors -- the lightly and heavily
taxed sectors. At times these are referred to as the corporate and non-corporate sectors due to the major role played by
the corporation income tax in causing the differential rates,
although his sectoral division does not exactly correspond to
the legal distinction between incorporated and unincorporated
enterprises. He assumes that each sector is perfectly competitive and employs tow factors of production, capital services and labor.

In order to estimate the efficiency loss due to the
differential taxation of the return to capital, Harberger applies a form of welfare analysis in the tradition of Marshal-
lian producer surplus. He assumes that the marginal product
schedules for each of his two sectors are linear as shown
below.

Output units are chosen so that both output prices are unity, and therefore these can be thought of as the value of the marginal product schedules as well. It is assumed that all prices other than the price of capital are unaffected by the presence of the differential capital income taxes and that the total amount of capital and labor are likewise constant. With these assumptions, the changes in the capital allocation can be used to generate a measure of the social waste imposed by the distortion. In the absence of any taxes, capital will allocate itself such that the rate of return is equalized across the two sectors (at $\bar{r}$ as shown above) and the capital endowment will be fully employed. Upon the imposition of a tax on capital income in sector X , the gross rate of return r_g in that sector must be such that the net rate of return r_n is equalized across the sectors and such that capital is again fully employed. The difference between r_g and r_n is, by definition, the tax T per unit of capital utilized in sector X .

On the graphs shown above, the area $ABEF$ can be interpreted as the value of the loss in output in sector X when K_X decreases from K_{X_o} to K_{X_1} upon the imposition of the tax. $GHIJ$, analogously, is the value of the increase in output in sector Y . Since we know that capital is fully employed both in the presence and absence of the tax, it must be true that $K_{X_o} - K_{X_1} = K_{Y_1} - K_{Y_o}$. The area $FECD$ represents the social loss of cost due to the distortionary tax and is simply $ABEF - GHIJ$. This loss can also be written as

$$\frac{1}{2}(r_g - \bar{r})(K_{X_o} - K_{X_1}) + \frac{1}{2}(\bar{r} - r_n)(K_{Y_1} - K_{Y_o}) = \frac{1}{2} T\Delta K_X \qquad (4)$$

where $\Delta K_X = K_{X_o} - K_{X_1} = K_{Y_1} - K_{Y_o}$ and $T = r_g - r_n$.

Now, in order to estimate the magnitude of this efficiency loss, a solution for ΔK_X is required. To solve for ΔK_X , Harberger abruptly changes models and uses a system of equations corresponding to the static two sector, two factor, general equilibrium model earlier developed by James Meade and Harry Johnson. Solving through a set of equations, the following result for ΔK_X is obtained:

$$\Delta K_X = K_X \cdot T \cdot \frac{-E\left[g_K S_X \dfrac{\frac{L_X}{X}}{L_Y} + f_K S_Y\right] - S_X S_Y f_L}{E\left(g_K - f_K\right)\left(\dfrac{\frac{K_X}{X}}{K_Y} - \dfrac{\frac{L_X}{X}}{L_Y}\right) - S_Y - S_X\left(f_L \dfrac{\frac{K_X}{X}}{K_Y} + f_K \dfrac{\frac{L_X}{X}}{L_Y}\right)} \qquad (5)$$

where

   E = price elasticity of demand for X

S_X (S_Y) = elasticity of factor substitution in sector X (Y)

f_K (g_K) = share of capital in sector X (Y)

f_L (g_L) = share of labor in sector X (Y) .

So, as one would expect, the capital shift depends on the various elasticities and factor intensities and is also proportional to T . This last point is emphasized as it later becomes important in that it implies that the social cost varies with the square of the tax rate.

He similarly solves this system of equations for the change in the net price of capital (i.e., $P_K = \bar{r} - r_n$) and obtains

$$\Delta P_K = T \cdot \frac{E f_K \left(\dfrac{K_X}{K_Y} - \dfrac{L_X}{L_Y}\right) + S_X \left(f_L \dfrac{K_X}{K_L} + f_K \dfrac{L_X}{L_Y}\right)}{E\left(g_K - f_K\right)\left(\dfrac{K_X}{K_Y} - \dfrac{L_X}{L_Y}\right) - S_Y - S_X \left(f_L \dfrac{K_X}{K_Y} + f_K \dfrac{L_X}{L_Y}\right)} \quad . \quad (6)$$

He uses this figure to evaluate the incidence of the burden reasoning that if $\Delta P_K = -TK_X / (K_X + K_Y)$ capital could be said to bear the full burden of the tax in that its gross return would be unchanges, while if $\Delta P_K = 0$, both input prices would be unaffected and thus, the share of national income going to capital or labor would also reamin constant.

Before examining the data which Harberger used to evaluate these two expressions, a few comments about the appropriateness of his techniques seems in order. First, for evaluating a distortion as large as that involved here, the local and linearization assumptions of the model seem inappropriate. For example, in solving for the efficiency loss formula (4), it has been assumed that the marginal product of capital varies linearly with the amount of capital for each of the sectors. At the same time, in the model used to solve for the capital shift, ΔK_X , constant returns to scale in production was assumed. These two assumptions are inconsistent for large changes since linear marginal product schedules imply a term quadratic in capital services in the production functions. Likewise assuming that output prices are unity both before and after the change is clearly a local

assumption. There are other shortcomings to the Harberger
analysis. First, it is restricted to a two sector, two factor model in which the total factor supplies are fixed. Second, there is only one consumer or market demand function in
the model and therefore the incidence issue can only be addressed with respect to the relative shares of capital and
labor. The issues relating to the perhaps more important
incidence, the impact on the distribution of income amongst
consumers, cannot be dealt with given Harberger's framework.
The algorithmic approach, at least in principle, suffers from
none of these shortcomings.

In order to evaluate the expressions (5) and (6), Harberger, in his 1966 article uses and supplements the data of
Rosenberg [6]. Taking as his unit of capital that amount
which earns one dollar net of taxes, he uses the following
table 1 to obtain data of K_X, K_Y, and T . Columns (1) and
(2) are drawn from a disaggregated study of Rosenberg [6].
Column (3) is meant to reflect the impact of the personal income tax on income from capital. Appealing to columns (4)
and (5), Harberger notes that total taxes on net income in
the "corporate" and "non-corporate" sectors average respectively 45 percent and 168 percent. Thus, he asserts that the
taxation of income from capital in the United States during
this period may be approximated by a general tax of 45 percent on all net income fram capital and an 85 percent surtax
on the net income from capital originating in the heavily
taxed sector (1.45 x 1.85 = 2.68).

Unfortunately, Harberger's data (reproduced here as
Table 1) contain a simple arithmetic mistake which greatly
affects his results. The entry superscripted d in column
(3) should be 5898 rather than 9945, and, with this

TABLE 1

Taxes on Income from Capital, by Major Sectors

(Annual Averages, 1953–1959, in millions of dollars)

	(1) Total Income* from Capital	(2) Property and Corp. Income Taxes	(3) Other Tax Adjustments	(4) Total Tax on Income from Capital	(5) Net Income from Capital
"Non-Corporate" Sector	26,873	6,639	1,724	8,363	18,510
Agriculture	7,481	1,302	927[a]	2,229	5,252
Housing	18,429	5,140	797[b]	5,937	12,492
Crude Oil and Gas	963	197	---[c]	197	766
"Corporate" Sector	52,399	22,907	9,945[d]	32,852	19,547
TOTAL	79,272	29,546	11,669	41,215	38,057

*"Income" (Rosenburg [6, p. 125] is defined as income from capital for non-financial industry and includes
(1) Corporate sector net income before corporate profits tax liability and property tax payments.
(2) For the unincorporated sector, the portion of the total income of the unincorporated enterprise that is a return on equity capital, plus property tax payments.
(3) Net monetary interest paid by businesses on borrowed capital in the form of debt obligations.
(4) Net rent paid by an industry to persons for the use of physical capital.
(5) Net realized capital gains by the corporate sector that are considered as income to an industry.

[a]Assumes a 15% effective tax on income from capital in agriculture after payment of property and corporate income taxes (i.e. (3) a = 15% of (1) − (2)).

TABLE 1 (continued)

Taxes on Income from Capital, by Major Sectors

(Annual Averages, 1953–1959, in millions of dollars)

---

[b] Assumes that 70% of income from capital in the housing sector is generated by owner-occupied housing, on which no personal income tax liability is incurred. It is assumed that the remaining 30% of capital income from housing is subjected to a 20% income tax rate after the deduction of property and corporate income taxes incurred (i.e., (3) b = 6% of (1) − (2)).

[c] Assumes personal tax offsets on account of oil depletion allowances and similar privileges offsets any taxes on dividends and capital gains in this sector.

[d] Assumes a 50% dividend distribution rate, and a "typical" effective tax rate of 40% on dividend income (i.e., d = 20% of (1) − (2)).

correction, the net income from capital in the "corporate" sector becomes 23,589 rather than 19,547 million dollars. The total tax on income from capital should be 28,805 million dollars. Table 2 contains the corrected factor input and capital tax data for both a two and twelve sector level of aggregation. The capital income and tax data are drawn from Rosenburg [6] as are the data of columns (1) and (2) of Table 1 and the "Other Tax Adjustments" column was constructed upon the assumptions listed in the footnotes to Table 1. The labor data are derived from National Income and Product Accounts. At the two sector level of aggregation, the main difference between the two tables is caused, of course, by the arithmetic mistake noted above. The surtax rate, which Harberger figures as 85 percent, is, according to Table 2, only 53 percent (1.45 x 1.53 = 2.22). The effect of the error is even greater than this since, as was pointed out earlier, the loss estimate varies as the square of the rate of distortionary taxation.

With the data of Table 1 augmented with (only approximately correct) labor data, Harberger evaluates ΔK_X and $-\frac{1}{2} T K_X$ for several different elasticities of factor substitution and for two different elasticities of demand for X , $-\frac{1}{7}$ and $-\frac{1}{14}$. The last two parameter values are derived from assumptions that the elasticity of substitution between X and Y is 1 and 1/2 respectively. His published results are (Harberger [4]) given in Table 3.

Using the data from Table 2 and simply plugging in the corrected numbers into Harberger's equations yields the following Table 4. As is clear, the efficiency loss estimates are lowered by a factor of up to four, now ranging from .42

TABLE 2
FACTOR PAYMENTS AND CAPITAL INCOME TAXES, BY MAJOR SECTORS

(Annual Averages, 1953-1959, in millions of dollars)

	(1) Total Return to Labor $ Million	(2) Total Return to Capital $ Million	(3) (1)+(2) Total Factor Return $ Million	(4) Property & Corporate Income Tax $ Million	(5) Other Capital Tax Adjustments $ Million	(6) (4)+(5) Total Capital Tax $ Million	(7) (2)-(6) Net Return to Capital $ Million	(8) (6)/(2) Capital Tax as a Fraction of Total Return to Capital	(9) (6)/(7) Capital Tax as a Fraction of Net Return to Capital
"Non-Corporate" Sector	17471	26878	44349	6639	1724	8363	18515	.3111455	.4516878
1 Agriculture	8800	7481	16281	1302	927	2229	5252	.2979548	.4244097
2 Real Estate	6869	18429	25298	5140	797	5937	12492	.3221552	.4752641
3 Crude Oil and Gas	1802	968	2770	197	--	197	771	.2035123	.2555123
"Corporate" Sector	199871	52394	252265	22907	5898	28805	23589	.5497765	1.2211220
4 Mining[1]	2528	688	3216	305	77	382	306	.5552325	1.2463660
5 Contract Construction	16670	1195	17865	435	152	587	608	.4912133	.9654605
6 Manufacturing[2]	79626	24665	104291	12488	2435	14923	9747	.6050273	1.5318209
7 Lumber & Wood Products	2426	718	3144	206	102	308	410	.4289693	.7512195
8 Petroleum & Wood Products	1846	3028	4874	770	452	1222	1806	.4035667	.6766334
9 Trade	43590	10897	54487	3493	1481	4974	5923	.4564559	.8397771
10 Transportation	14078	2683	16761	1230	291	1521	1162	.5669027	1.3089500
11 Communication and Public Utilities	7394	6489	13883	3290	640	3930	2559	.6056403	1.5357561
12 Services	31713	2031	33744	690	268	598	1073	.4716888	.8928238
TOTAL	217342	79272	296614	29546	7622	37168	42104	.4688666	.8827664

[1] Other than Crude Oil and Gas

[2] Other than Lumber and Wood Products and Petroleum and Coal Products

TABLE 3

Harberger's Estimates of Efficiency Cost of Existing

Taxes on Income from Capital in the U.S.:   1953–1959

S_X	S_Y	E	ΔK_X (billions of units)	$-\frac{1}{2} T\Delta K_X$ ($ billion)
-1	-1	$-\frac{1}{7}$	-6.9	2.9
-1	$-\frac{1}{2}$	$-\frac{1}{7}$	-5.9	2.5
$-\frac{1}{2}$	-1	$-\frac{1}{7}$	-5.2	2.2
$-\frac{1}{2}$	$-\frac{1}{2}$	$-\frac{1}{7}$	-4.8	2.0
-1	0	$-\frac{1}{7}$	-4.7	2.0
$-\frac{1}{2}$	0	$-\frac{1}{7}$	-3.9	1.7
-1	-1	$-\frac{1}{14}$	-5.3	2.3
-1	$-\frac{1}{2}$	$-\frac{1}{14}$	-4.2	1.8
$-\frac{1}{2}$	-1	$-\frac{1}{14}$	-4.1	1.7
$-\frac{1}{2}$	$-\frac{1}{2}$	$-\frac{1}{14}$	-3.5	1.5
-1	0	$-\frac{1}{14}$	-5.0	2.1
$-\frac{1}{2}$	0	$-\frac{1}{14}$	-2.4	1.0

TABLE 4

Revised Efficiency Cost Estimates Using Harberger's Model

1953 - 1959

S_X	S_Y	E	ΔK_X (billions of units)	$-\frac{1}{2} T\Delta K_X$ ($ billion)
-1	-1	$-\frac{1}{7}$	-4.63	1.22
-1	$-\frac{1}{2}$	$-\frac{1}{7}$	-4.01	1.06
$-\frac{1}{2}$	-1	$-\frac{1}{7}$	-3.56	$.94$
$-\frac{1}{2}$	$-\frac{1}{2}$	$-\frac{1}{7}$	-3.25	$.86$
-1	0	$-\frac{1}{7}$	-3.22	$.85$
$-\frac{1}{2}$	0	$-\frac{1}{7}$	-2.80	$.73$
-1	-1	$-\frac{1}{14}$	-3.52	$.93$
-1	$-\frac{1}{2}$	$-\frac{1}{14}$	-2.74	$.72$
$-\frac{1}{2}$	-1	$-\frac{1}{14}$	-2.78	$.73$
$-\frac{1}{2}$	$-\frac{1}{2}$	$-\frac{1}{14}$	-2.32	$.61$
-1	0	$-\frac{1}{14}$	-1.75	$.46$
$-\frac{1}{2}$	0	$-\frac{1}{14}$	-1.61	$.42$

to 1.22 billion dollars rather than the earlier reported 1.0
to 2.9 billion dollars. It is these lowered revised numbers
which should be compared with the results of the algorithmic
technique.

Using the revised data of Table 2, I have recomputed
the efficiency costs of the distortionary taxation of capital
income and briefly examined the incidence question as well.
This has been done for both a two sector and twelve sector
level of disaggregation to gain some feeling for the sensitivity of the results to aggregation. Where possible, the assumptions of Harberger have been followed, although the models, of course, are not strictly comparable. Production
functions were assumed to be of the CES form, while both CES
and Cobb-Douglas utility functions were analyzed for consumers. Two classes of consumers were incorporated, one representing the top ten percent of the income recipients in the
U.S., the other the bottom ninety percent. The model was
solved both with and without the inclusion of a labor-leisure
choice.

The technique was, of course, to compare the current
(distorted) equilibrium with the equilibrium which would result in the absence of the distortionary taxes. The flat
capital tax rate required to match the distortionary taxes
was computed. As this is probably not the proper forum for
a complete listing of the extensive cases examined, only the
ones most directly comparable to Harberger's will be discussed. In particular, if the two consumers have demand functions derived from Cobb-Douglas utility functions and assuming a fixed total amount of capital and labor, the algorithmic approach yields results summarized in Table 5 for
three cases closely comparable to Harberger's. From Table 5,

TABLE 5

Two Sector Results Using Algorithmic Technique

	Observed Equilibrium	New Calculated Equilibrium 1	2	3
S_X		−1.0	−1.0	−1/2
S_Y		−1.0	−1/2	−1/2
P_X	1.00000	.96626	.97136	.95607
P_Y	1.00000	1.17068	1.19455	1.13088
P_L	1.00000	1.00000	1.00000	1.00000
P_K	1.00000	1.29695	1.33027	1.22027
R	37.168	24.665	25.299	23.207
X	252.265	261.074	261.495	259.666
Y	44.349	37.883	37.374	38.614
K_X	23.589	27.828	27.318	26.585
K_Y	18.515	14.276	14.786	15.519
L_X	199.871	199.871	201.250	201.165
L_Y	17.471	17.471	16.092	16.177
$\left(1 + \dfrac{\Delta P_K K}{R}\right)$		.00000	−.17158	.33570
ΔNNP		2.344	2.255	1.666
$\dfrac{\Delta NNP}{\Delta R}$		.18747	.18999	.11933

two things are clear. One, the algorithmic technique, even
for the two sector case, provides a more complete description
of the new equilibrium. Secondly, the loss estimates, here
reported as the change in net national product evaluated at
current, observed prices, are just about double what the
Harberger model predicts for the corresponding cases (rows 1,
2, and 4 of Table 4). These losses, while small relative to
GNP, amount to almost twenty percent of the revenue of the
distortionary taxes.

On further examining Table 5, one sees that most parameters behave as one would expect. With the removal of the
distortionary taxes, the price of X falls, the price of Y
rises as does the net price of capital. Sector X becomes
more capital intensive, while Y becomes more labor inten-
sive. The term $1 + \dfrac{\Delta P_K K}{\Delta R}$ attempts to measure the incidence of the tax on the functional distribution of income.
If it were zero, capital could be said to be bearing the full
burden while if positive, labor would be sharing in the burden. As the taxed sector is relatively labor intensive, it
is possible for a distortionary tax on capital in that sector to adversely affect labor as is witnessed by the results
for the $S_X = S_Y = -\dfrac{1}{2}$ case in Table 5. For a more complete
description of capital income tax results see Shoven [8,10].

The few results reported here do give, I feel, an accurate impression that the answers from the algorithmic approach differ substantially although not, perhaps, dramatically from those of the earlier methodology. Upon disaggregating to twelve production sectors, the loss estimates increased another twenty-five percent or so. They then go as
high as $3.3 billion for the case where Harberger's approach

yielded $1.22 billion. This amounts to just about twenty-
five percent of the yield of the distortionary portion of the
taxes.

It is my feeling that the results of the two techniques
are significantly different and that this algorithmic approach is by far the more appropriate methodology. As will
be indicated in the next section, I feel the potential for
this approach is very great indeed, although a good deal of
further work on the techniques themselves is probably necessary.

4. Further Research Possibilities

With the exception of John Whalley's thesis evaluating
the April 1973 British tax reforms, the capital income tax
application just discussed is the only tax analysis which has
been approached with the algorithmic methodology. It seems
clear to me that there are many other promising applications
and I will mention some of the most obvious in this section.

First, in the capital income tax study just described,
only that single distortion was evaluated. One of the advantages of this type of technique is that more than one distortion can be simultaneously modeled. Thus, a natural extension will involve embedding the capital income tax structure
into the entire U.S. tax system. That is, at the least, the
federal income tax and the social security system will be included. Whalley's study did include the entire British tax
system, thus indicating that this is a feasible, although not
a simple, undertaking. Once the entire tax system was modeled, the effects of changing one part of it on the other
components of the tax system could be evaluated. Also, such

policy possibilities as the integration of the corporation
and the personal income tax systems, financing social security out of increased income taxes, introducing a negative
income tax, or instituting a value added tax as a partial replacement for property taxes could be evaluated. Each of
these are relevant public policy alternatives and involve
very large, non-marginal adjustments. A slightly different
application, not involving taxes directly, would be to evaluate wage-subsidy programs as alternatives to either negative
income taxation or the present welfare system. The impact of
labor unions could likewise be evaluated. The algorithmic
approach seems appropriate for the analysis of each of these
large change proposals.

In addition to these potential applications, additional
work is needed in developing a more detailed model of both
the consumer and producer sides of the economy. There seems
to be little problem in disaggregating market demand into
many separate demand functions for different consumer classes. On the production side, however, additional richness
comes less easily. Both an activity analysis approach and
the standard (say, CES or Cobb-Douglas) production functions
seem less than satisfactory, the former because of the lack
of substitutability of factors, and the latter due to the
problems encountered with many, including intermediate, inputs. In Shoven [8], the Sato-Uzawa extension to CES production of each commodity may use any or all other commodities
as inputs, yet, at the same time, one wants to allow continuous input substitution. The recent work of Diewert [1] has
produced just such a description of technology and the incorporation of his technology into a framework such as that described above seems feasible. Diewert is currently involved

in econometrically estimating such a production technology
for Canada and its inclusion in a model of the U.S. seems
most desirable.

Returning to applications, perhaps the most important
are those involving several countries. John Whalley is currently working on an evaluation of European fiscal harmonization while he and I together have begun a project analyzing
the impact of tariffs on Japanese-American trade. The effects of import quotas or of a crude oil embargo such as recently experienced likewise seem to be problems calling for
an approach such as this.

Of course, progress on the computational techniques
themselves in computational speed, etc. will allow the evaluation of larger, more complex models. Dynamic extensions of
the currently static models will then be feasible, permitting
a more satisfactory analysis of such phenomena as capital
gains. These larger, detailed, dynamic models would require
large amounts of data, but perhaps progress on its availability will be made simultaneously. All in all, then, I see the
application of fixed point algorithms as an extremely promising area for further research. With this research, these
techniques will become an important tool of economic analysis.

REFERENCES

[1]   Diewert, W. E., "An Application of the Shepard Duality
      Theorem: A Generalized Leontief Production Function",
      The Journal of Political Economy, Vol. 79, pp. 481-507,
      1971.

[2]   Harberger, A. C., "The Corporation Income Tax: An Em-
      pirical Appraisal", in Tax Revision Compendium, House
      Committee on Ways and Means, 86th Congress, 1st ses-
      sion, Vol. 1, pp. 231-240, 1959.

[3]   Harberger, A. C., "The Incidence of the Corporation In-
      come Tax", Journal of Political Economy, Vol. 70, pp.
      215-240, 1962.

[4]   Harberger, A. C., "Efficiency Effects of Taxes on In-
      come from Capital", in M. Kryzaniak, eds., Effects of
      the Corporation Income Tax, Detroit, Wayne State Press,
      1966.

[5]   Robinson, S. M., "Extension of Newton's Method to Mixed
      Systems of Nonlinear Equations and Inequalities",
      mimeo, 1972.

[6]   Rosenburg, L. G., "Taxation of Income from Capital, by
      Industry Group", in A. C. Harberger and M. J. Bailey,
      eds., The Taxation of Income from Capital, Washington,
      D.C., The Brookings Institution, 1969.

[7]   Scarf, H. E., with the collaboration of T. Hansen, The
      Computation of Economic Equilibria, New Haven, Yale
      University Press, 1973.

[8]   Shoven, J. B., General Equilibrium with Taxes: Exis-
      tence, Computation, and a Capital Income Taxation Ap-
      plication, unpublished Ph.D. thesis, Yale University,
      1973.

[9]   Shoven, J. B., "A Proof of the Existence of a General
      Equilibrium with Ad Valorum Commodity Taxes", Journal
      of Economic Theory, Vol. 8, No. 1, pp. 1-25, 1974.

[10] Shoven, J. B., "The Incidence and Efficiency Effects of
      Taxes on Income from Capital", mimeo, 1974.

[11] Shoven, J. B., and Whalley, J., "A General Equilibrium
      Calculation of the Effects of Differential Taxation of
      Income from Capital in the U.S.", Journal of Public
      Economics, Vol. 1, No. 3, pp. 281-321, 1972.

[12] Shoven, J. B., and Whalley, J., "General Equilibrium
      with Taxes: A Computational Procedure and an Exis-
      tence Proof", The Review of Economic Studies, Vol. XL,
      No. 4, pp. 475-489, 1973.

[13] Shoven, J. B., and Whalley, J., "On the Computation of
      Competitive Equilibrium on International Markets with
      Tariffs", forthcoming in Journal of International
      Economics, 1974.

[14] Shoven, J. B., and Whalley, J., "Computing Competitive
      Equilibria Under Tax Replacement Schemes", unpublished
      mimeo, 1974.

[15] Whalley, J., "A Numerical Assessment of the April 1973
      Tax Changes in the United Kingdom", unpublished Ph.D.
      thesis, Yale University, 1973.

"Fiscal Harmonization in the EEC;

Some Preliminary Findings of Fixed Point Calculations"[1]

John Whalley

ABSTRACT

This paper examines the process of harmonization on a common output tax system among EEC member states within a general equilibrium framework. For each of the economies in the EEC the tax system is, in turn, replaced by each of the tax systems prevailing in other member states and the new competitive equilibria are computed. Data for 1965 common to each of the original member states are used which involve computational problems in high dimensional space (61 dimensions). Using features of the model a lower dimension can be used for curcial components of the calculations. It is found

---

[1]I am grateful to the Houblon-Norman Fund for their support during this research, and also to H. Krijnse Locker and P. Le Grontec of the Statistical Office of the European Communities for their help in providing and interpreting data.

that Newton methods designed to refine the approximation from simplicial subdivision methods can be applied in their own right as global methods. Even though there is no guarantee of convergence this problem has not been encountered and execution is very fast (less than 6 seconds on a CDC 7600). Results indicate the French tax system to be least desirable and the Dutch to be most desirable for all the EEC member countries. Substantial gains are, however, forgone by not harmonizing on a non-distortionary tax system.

1. Introduction

This paper reports some preliminary findings from the
application of fixed point computational methods to assess-
ments of the impacts of fiscal harmonization procedures in
the EEC. As such, this represents a further application of
computational methods for the calculation of fixed points to
policy issues connected with the analysis of tax reforms. A
formulation used in the earlier applications of Shoven and
Whalley [6], and Whalley [10] is followed except for a more
complex specification of the production side of the economy
to include intermediate production. In addition, problems
of large dimensionality have been analyzed involving simplices higher than sixty dimensions. The increased dimensionality made possible by the disaggregation of the tax and
production side data enables more detail within the tax system to be taken into account than in earlier work. This is
of considerable benefit when one is concerned with assessing
discriminatory aspects of tax systems, and the possibilities
for improvement in the operation of the economy. Results obtained from this model are reported and their relevance to
the fiscal harmonization debate within the EEC discussed;
computational experience is also outlined.

2. Taxation in General Equilibrium Models

Shoven and Whalley [7] have recently formulated a
general equilibrium model of the operation of tax systems in

economies for use in the analysis of tax reforms. In a model of the economy with N commodities, vectors of ad valorem
tax rates on each commodity are assigned to economic agents.[1]
Consumers acquiring commodities for use in consumption activities must pay taxes on these purchases; producers purchasing
inputs or selling outputs must also pay taxes. The government acts as a collection agency for tax receipts which may
be distributed between customers (transfer payments) or re-
tained (in part or wholly) by the government to finance ex-
penditures on public authority goods and services.

With this model, competitive equilibria in the presence
of taxation are examined. Such states are characterized by
two sets of conditions (i) equalities between quantities demanded and supplied (including disposals) for each commodity
at an equilibrium vector of market prices (ii) conditions
that at the equilibrium vector of market prices no producer
does any better than break-even (net of producer taxes).
Corresponding to each equilibrium is thus a vector of equilibrium market prices, a vector of equilibrium quantities of
each commodity, and an equilibrium value of tax receipts.

These models also embody a number of assumptions which
enable proofs of the existence of such equilibria to be

---

[1]Increasingly, certain kinds of taxes are being administered on a transactions basis and these kinds of taxes
can cause some complications with this framework. Gift and
estate taxes, for instance, are of this form. Value-added
tax systems are also administered on a transactions basis,
but as they have tax rates on transactions which differ according to the commodities rather than the individuals involved they can be accommodated within this framework. This
involves a convertion of the legal V.A.T. transactions tax
rates into effective commodity tax rates.

obtained and fixed points to be computed. On the production
side of the economy constant returns to scale is assumed[1]
along with boundedness conditions assuring that if the endowments of goods in the economy are finite then it is not possible to produce infinite amounts of any good. On the demand
side of the economy market demand functions which are non-
negative, continuous, and homogeneous of degree zero in all
market prices and tax revenue[2] are assumed. Market demand
functions are also assumed to satisfy the condition that the
value of demands (at any vector of prices and tax revenue)
plus consumer tax revenue raised must equal the value of endowments plus the tax revenue distributed. This is a property known as Walras Law which may be interpreted as a requirement that all consumer incomes are spent on commodities
and consumer tax payments. Purchases by individuals are financed by each person selling his endowment of goods (assumed privately owned); the proceeds from such a sale together with transfers received from the government determining the consumer's income.

The existence of competitive equilibria in these models has been investigated by Shoven [8], and Shoven and
Whalley [7] among others. Procedures for the computation of
such equilibria, modelled on Scarf's fixed point algorithm

---

[1]For the purpose of proofs of the existence of equilibria, decreasing returns to scale can be considered. See
Shoven [8].

[2]The notion is that quantities demanded are dependent only on relative prices, and tax revenue itself is
treated as a relative price. A doubling of all prices and
tax revenue doubles the income of any consumer but also
doubles the price per unit of any commodity. Physical quantities demanded are thus assumed to be unchanged.

[3], [4], [5], have been used by Shoven and Whalley [7] and
Whalley [10]. These procedures form the basis of the application reported on here, and which is explained in more detail
in the next section.

3. Fiscal Harmonization in the EEC; A General Equilibrium
    Approach

    The present nine member countries of the EEC are cur-
rently engaged in a process of aligning their domestic taxation arrangements towards a single common system[1] which will
eventually prevail in each country. The origins of this process lie in the signing of the Treaty of Rome in 1958 by the
original six members of the EEC which carried with it a com-
mitment to remove barriers to competition between states.
The first stage in this procedure involved the abolition of
tariff barriers on inter-country trade, but it was soon recognized that a harmonization on common taxation arrangements
within countries would also be necessary to meet the commit-
ment to free trade. Domestic taxes on consumption of various
commodities differ from country to country and at present any
one country faces different tax rates on its exports depending upon which country they are sold to. This distorts
international trade patterns from what would be required for
an efficient allocation of resources on a worldwide basis.

---

[1]There are, of course, problems of definition with
the term "common system". Thus far, all EEC countries have
value-added tax (V.A.T.) systems but tax rates and tax bases
vary from country to country; hence this first stage in the
harmonization process has involved adopting a common nomen-
clature and administration rather than a common set of commodity tax rates.

The harmonization process in the EEC has so far been focused on a sequence of partial harmonizations. One portion of a target common tax system replaces a group of existing taxes in each country, followed at a later date by a common tax rate [1] for this portion. The process then moves on to further taxes. So far, only harmonization of output tax systems has been accomplished[2] with harmonization of output tax bases and tax rates still some way in the future. The beginnings of the process of harmonization of corporate taxes are only now being detected.

Although the motivation for fiscal harmonization among EEC member countries lies in a desire to partially remove distortions of international trade patterns, the fiscal harmonization procedures will, of course, also have a considerable domestic impact in each of the countries. It is this

---

[1] This would correspond to all EEC countries having common VAT rates and a common tax base.

[2] A directive of the Council of Ministers in 1967 called for the adoption of value-added tax systems in all member countries by 1970. At that time only France had a VAT system, but W. Germany complied in 1970, Italy in 1973, and the UK on joining in 1973. A harmonization of VAT bases and rates is envisaged by 1978. To illustrate the substantial differences remaining, the following points may be noted: France has four different rates of tax, W. Germany has two, but the UK has only one; the UK does not include food in the tax base whereas most other EEC countries do. Discussions have recently been taking place in Brussels on the target common corporate tax system (classical, imputation, or split rate), but no directive has yet been issued. The harmonization process is thus unlikely to be complete (if ever) for decades; no movement towards harmonization has been made for the majority of taxes, income taxes, property taxes, estate and gift taxes, excise taxes, social security taxes and other fiscal instruments.

domestic impact which is the focus of the present report.
What is the impact on, say, the German economy of adopting
the French tax system? Do the Germans gain or lose, and what
are the magnitudes involved? If one of the existing national
tax systems is to be adopted by all the member states, which
is the most preferable system? How much could be gained by
a harmonization instead, on a non-distortionary tax system?

The impact on international trade flows is a subject of
considerable interest and importance but is not a matter capable of useful analysis with the model used here given present data availability. To adequately capture the features
of taxation arrangements in these economies a considerable
degree of disaggregation by industry is needed; trade flow
data by commodity between each pair of countries is not easily obtained in a way which is compatible with other data
used here. This would have to be aggregated from extremely
detailed information.[1]

Thus a series of calculations has been performed for
each country. A static general equilibrium model of each of
the economies involved has been constructed with the domestic
tax system introduced into the model. For each of the countries the domestic tax system is then, in turn, replaced by
each of the tax systems operating in other EEC countries and
the impacts assessed. Due to the empirical difficulties of

---

[1]The fact that it is data availability rather than
dimensionality of the problem that is a constraint in this
application is a point worth some emphasis. Each country has
over fifty commodities and a true international trade flow
application would involve over 300 commodities. Given the
structure which is placed on the production side of the economy, problems of this dimensionality could probably be han-
dled computationally if reliable data were available.

adequately specifying the whole tax system of each of the
countries involved, only the output tax systems within each
country are considered.[1]

Each economy is represented by a separate general equilibrium model involving the production side of the economy,
the demand side of the economy, the tax system and the government budget, and the external sector (foreign trade).

On the production side of the economy a distinction is
made between outputs and factors of production. Two factors
of production (capital and labour) are considered in each
economy; these may be substituted one for the other in the
production process for any output. Production of any output
also requires as inputs some of the other outputs (intermediate production). Each economy is treated as having endowments of factors of production, $\bar{K}$ and $\bar{L}$, and zero initial endowments of outputs. Endowments of factors are privately owned; factors of production are heterogeneous across
economies (French and German capital are treated as being
qualitatively different).

---

[1]This procedure of focusing solely on the output
taxes in each country seems reasonable as harmonization on a
common output tax system corresponds to that component of
the fiscal harmonization process where the procedures are
most advanced. As will be made clear later, the output tax
system in the model does not correspond solely to the VAT
and turnover tax systems which have been the object of recent
changes. The system includes all ESA (European System of
Accounts) "producer taxes" (broadly, VAT, turnover taxes, and
excises; but not corporate or property taxes). This involves
a simplification for each economy as the interaction of each
of these output tax systems with the rest of the tax system
in each economy is ignored.

The production processes are specified in more detail as follows: There are N outputs in each economy and to produce any output the other (N-1) may be required as inputs[1] in the production process. The term a_{ij} represents the amount of the jth good needed to produce one physical unit of the ith output; where $a_{ij} \leq 0$, $j \neq i$; $a_{ii} = 1$,[2] i,j = 1,...,N . These "activities" may be operated at any non-negative level. In addition, factor inputs (capital and labour) are also needed in any production process which may be substituted one for the other according to a continuous constant returns to scale function. This function also determines the level of production, S_i , of any output (the scale of operation). A convenient representation of these scale functions which is used here is the constant elasticity of substitution function

$$S_i = \gamma_i \left[\delta_i K_i^{-\rho_i} + (1-\delta_i) L_i^{-\rho_i} \right]^{\frac{-1}{\rho_i}} , \quad (i=1,...,N) \quad (3.1)$$

where $\sigma_i = 1/(1 + \rho_i)$ is the elasticity of substitution between factor inputs in the production process for the ith output, δ_i gives the "weighting" parameters between factor inputs, and γ_i is a parameter defining units of measurement for the ith output.

---

[1] Thus steel is needed as an input in the chemical industry, and chemicals as input in the steel industry.

[2] The amount of the ith output needed to produce itself has thus been netted out.

In addition to the functions (3.1), there is also a listing of vectors, one for each industry, giving the intermediate production requirements. These vectors $(A^i)' = (a_{i1}, \ldots, a_{iN})$ are arranged in the $N \times N$ matrix A.

$$
\begin{array}{c}
(A^1) \ldots \ldots (A^i) \ldots \ldots (A^N) \\
A = \begin{bmatrix}
1 & , \ldots \ldots \ldots \ldots \ldots a_{N1} \\
a_{12} & a_{N2} \\
\cdot & \cdot \\
\cdot & \cdot \\
\cdot & \cdot \\
\cdot & \cdot \\
\cdot & \cdot \\
a_{1,N-1} & a_{N,N-1} \\
a_{1N} & , \ldots \ldots \ldots \ldots 1
\end{bmatrix}
\end{array}
\qquad (3.2)
$$

Thus production processes for each output (industry) may be represented as in (3.3) by combining (3.1) and (3.2),

$$
(A^i)' \cdot S_i = (A^i)' \cdot \gamma_i \left[\delta_i K_i^{-\rho_i} + (1-\delta_i) L_i^{-\rho_i} \right]^{\frac{-1}{\rho_i}} ,
\qquad (3.3)
$$

$$
(i=1,\ldots,N)
$$

This description of the production side of the economy may be viewed as a partition of the commodities in the economy into those which are substitutable one for the other and those which are not. The reasons for adopting this particular formulation are worth mentioning. Capital and labour represent "primary" factor inputs into the productive activities of various industries, and the vectors $(A^i)'$ capture intermediate production requirements. This is a convenient analytical formulation for the application of national accounts data to empirical problems of this form. As such this extends the traditional two sector formulation of simple general equilibrium models used in the analysis of the impact of factor market distortions.[1]

On the demand side of the economy C groups of consumers are identified intended to capture groups of income recipients running schematically from rich to poor.[2] Each group is thought of as generating demands for commodities from the maximization of a utility function also of constant elasticity form subject to the group's budget constraint,

$$
U_c = \left[\sum_{i=1}^{N} (b_{ci})^{\frac{1}{Z_c}} (X_{ci})^{\frac{Z_c-1}{Z_c}} \right]^{\frac{Z_c}{Z_c-1}}, \quad (c=1,\ldots,C) \quad (3.4)
$$

The term X_{ci} represents the consumption of the ith commodity by the cth consumer, Z_c is the elasticity of substitution between the X_{ci} in the utility generating

---

[1]Distortionary tax systems are perceived to have a different impact in models with intermediate production than in models where it is absent. The early work in this area of Johnson [2] and Harberger [1] indicate small efficiency impacts of discriminatory factor taxes. These results are likely to be changed somewhat by the introduction of intermediate production into the model.

[2]In the preliminary findings reported in this paper only one group of consumers is considered in each country. Extensions of the empirical application of this model will involve more than one grouping as in Whalley [10].

process, and b_{ci} is the "weight" attached to the ith good. Faced with a vector of market prices $P' = (P_1,\ldots,$ $P_N, P_K, P_L)$, and given the cth group's receipt of tax revenue from the government; I_c the cth group's income is determined at these prices and maximization of (3.4) subject to the budget constraint for this group yields the demand functions

$$X_{ci} = \frac{b_{ci}\, I_c}{\left[P_i(1-t_i^c)\right]^{Z_c} \sum_{j=1}^{Z_c} b_{cj} \left[P_j(1-t_j^c)\right]^{1-Z_c}},$$

$$(i=1,\ldots,N)$$
$$(c=1,\ldots,C) \qquad (3.5)$$

where the P_i are prices net of consumer taxes (producer market prices).

The tax system introduced into the model assigns a vector of ad valorem tax rates to each group of consumers in each of the EEC countries $(t^c)' = (t_1^c,\ldots,t_N^c)$. These taxes are paid on purchases of each of the N commodities. For any vector of market prices $P' = (P_1,\ldots,P_N)$ the tax revenue which is received by the government is

$$R = \sum_{c=1}^{C} \sum_{i=1}^{N} t_i^c \, P_i X_{ci} . \qquad (3.6)$$

The expenditure side of the government budget is assumed to be solely redistributive, dividing up any given tax revenue R among the consumers in the economy (including the government itself). A scheme of fixed proportional

division is used although a more complex formulation in which
proportional shares vary with market prices could be used if
data were available to parameterize such functions.

Lastly, each of the economies is assumed to have an
external sector representing trade with the rest of the
world (including other EEC countries). One other country
ROW[1] (rest of the world) is considered which is endowed with
capital and labour (assumed heterogeneous from domestic factors of production) which are used to produce a single output
which is exported from ROW. This is produced according to a
function as in (3.1). ROW then imports some of the domestic
outputs, imports being financed by sales of exports.

Thus, in total, the model used for each economy has
N + 5 commodities; N domestic outputs; 1 output in ROW;
K^D, L^D as domestic factors of production; and K^{ROW}, L^{ROW} as
factors of production in ROW. An equilibrium price vector
will include the equilibrium value of tax receipts and will
be (N+6) dimensional $(P_1, \ldots, P_N, P_{ROW}, P_K^D, P_L^D, P_K^{ROW}, P_L^{ROW}, R)$.

4. Computation of Competitive Equilibria in the EEC Tax Model

For the model presented in section 3 competitive equilibria for each economy in the presence of taxation are examined. As mentioned earlier a competitive equilibrium with
taxes is characterized by a vector of prices and tax revenue,
and a vector of scales of operation for each industry such
that the quantity demanded equals that supplied for each

---

[1]ROW is assumed to have no tax system.

commodity and no producer in the economy does any better than break even. Let $X_i(P,R)$ be the market demand for the ith commodity and $\bar{K}^D, \bar{L}^D, \bar{K}^{ROW}$, and $\bar{L}^{ROW}$ be the endowments of the factors of production in each economy. Then an equilibrium for each EEC country is characterized by a vector (P,R) and a vector S such that

$$X_i(P,R) - \sum_{j=1}^{N} a_{ji} S_j \leq 0 \qquad (i = 1,\ldots,N) \qquad (4.1)$$

$$X_{N+1}(P,R) - S_{N+1} \leq 0 \qquad (4.2)$$

$$\sum_{i=1}^{N} K_i^D \leq \bar{K}^D \qquad (4.3)$$

$$\sum_{i=1}^{N} L_i^D \leq \bar{L}^D \qquad (4.4)$$

$$K_{N+1}^{ROW} \leq \bar{K}^{ROW} \qquad (4.5)$$

$$L_{N+1}^{ROW} \leq \bar{L}^{ROW} \qquad (4.6)$$

$$P_i S_i - \sum_{j=1}^{N} P_j a_{ij} S_i - P_K^D K_i^D - P_L^D L_i^D \leq 0 \qquad (i=1,\ldots,N) \qquad (4.7)$$

$$P_{N+1} S_{N+1} - P_K^{ROW} K_{N+1}^{ROW} - P_L^{ROW} K_{N+1}^{ROW} \leq 0 \qquad (4.8)$$

The vector P is of dimension $N+5$, R is single dimensional, and S is of dimension $N+1$.

To compute these equilibria, termination methods developed to refine the approximation from simplicial subdivision methods such as those due to Scarf [3, 4, 5] can be used in their own right. It should be pointed out that these methods do not guarantee that a solution will be found as the non-cycling argument of simplicial approximation methods is no

longer available. These methods thus represent a form of
Gauss-Newton iterative procedure. They are much quicker than
simplicial subdivision methods and no cycling or non-conver-
gence problems have been encountered in this particular application. The methods used involve a process of improvement by solution of a sequence of linear programming problems.[1] A set of vectors which define a region of search on
a unit simplex are constructed around an initial guess (usu-
ally the mid point $(\frac{1}{K}, \ldots, \frac{1}{K})$ where K is the dimension-
sionality of the simplex).

To each vector correspond prices and a value of tax
revenue for which excess demands are computed and a linear
programming problem set up to choose weights (which may be
negative) for these vectors which come closest to meeting
the conditions of a competitive equilibrium under certain
linearization assumptions. The solution to this linear programming problem yields the next initial guess on the simplex which will not be the actual answer as the linearization assumptions will not hold. The process is then repeated, but if the solution to the previous linear programming problem is interior to the area of search, the new area
is correspondingly reduced. As the area of search is reduced the linearization assumptions will hold to a better
and better approximation until a solution within the desired
tolerance limits is found.

The equilibria which have been computed in this application lie on 61 dimensional simplices but with large

---

[1]These routines are derivatives of those developed
by J. B. Shoven and the original routine is explained in
some detail in [6] Appendix A and also in [9].

reductions in the dimensionality of the linear programming problems solved it is possible to reduce execution time to very small magnitudes. In all the cases examined in the paper it is anticipated that at a competitive equilibrium each industry will have positive net production. From the cost minimization problem for each industry the set of output prices which guarantee that zero profitability conditions will hold may be found and used in the determination of market demands. This method of working in a lower dimensional space and generating the N output prices as the procedure operates allows the removal of 2N rows and N columns from the constraint matrix of each linear programming problem which is solved.[1] (These problems will be explained below.) Moreover, given the constant returns to scale assumption, there is some scale indeterminacy in choosing appropriate levels at which to operate each cost minimizing activity when the supply response of each industry to any vector of prices is determined. By adopting the procedure of using output demands as scales of operation for each industry, the excess demands for outputs are equated to zero at each price vector removing a further N rows from the constraint matrix.[2]

---

[1] See the solution procedure in Shoven and Whalley [6] where the reduction in the dimensionality of the space is not used.

[2] For the problems examined in the text, the constraint matrix involved is reduced by these operations from (173 x 234) to (6 x 13). On a 61 dimensional simplex, solutions are typically often found within 20 linear programming problems taking about 15 seconds of execution time on a CDC 7600, giving all excess demands to less than 0.000000002. This is for cases where demands for some commodities are as large as 100,000.0.

452   JOHN WHALLEY

These computational methods may be described more explicitly as follows. Let $\tilde{P}^j_i$ be the ith component of a vector $\tilde{P}^j$ on a (N+6) dimensional simplex.[1] Using the techniques described above the search across this simplex can be restricted to the dimensions (N+2) to (N+6)[2]. Then with $(\tilde{P}^j_{N+2},\ldots,\tilde{P}^j_{N+6})$ as a point estimate on this lower dimensional subsimplex a set of 5 vectors surrounding it is generated which together define a region of search on this 5 dimensional subsimplex.

$$
P^1 = \begin{bmatrix} \tilde{P}^j_{N+2}\,(1-\tilde{\Delta})\,/\,D_{N+2} \\ \tilde{P}^j_{N+3}\,/\,D_{N+2} \\ \cdot \\ \cdot \\ \cdot \\ \cdot \\ \tilde{P}^j_{N+6}\,/\,D_{N+2} \end{bmatrix} ;\ldots ; P^5 = \begin{bmatrix} \tilde{P}^j_{N+2}\,/\,D_{N+6} \\ \tilde{P}^j_{N+3}\,/\,D_{N+6} \\ \cdot \\ \cdot \\ \cdot \\ \tilde{P}^j_{N+6}\,(1-\tilde{\Delta})\,/\,D_{N+6} \end{bmatrix} \quad (4.9)
$$

where $\tilde{\Delta}$ is a small number (say 0.04) and

$$
D_i = 1 - \tilde{\Delta}\tilde{P}^j_i \,/\, \sum_{k=N+2}^{N+6} \tilde{P}^j_k , \quad (i=N+2,\ldots,N+6) . \quad (4.10)
$$

---

[1] Note that the tax revenue term R has now been included in the vectors P^j as the (N+6)th entry.

[2] The space of the factors of production and tax revenue. There are N domestic outputs and 1 ROW output, (N+1) in all.

Corresponding to each of the 5 vectors in (4.9) a vector of excess demands for the four factors of production and tax revenue is calculated, assuming that excess demands for all (N+1) outputs are zero and that zero profitability conditions hold in each industry. This is done in the following manner; first the cost minimizing combination of substitutable factors of production in each industry needed to produce 1 unit of each of the (N+1) outputs is calculated. This involves solving the problems

$$\min \quad P_K^D K_i^D + P_L^D L_i^D$$

$$\text{subject to} \quad \gamma_i \left[\delta_i (K_i^D)^{-\rho_i} + (1-\delta_i)(L_i^D)^{-\rho_i} \right]^{-\frac{1}{\rho_i}} \leq 1$$
$$(i=1,\ldots,N)$$

and
$$\tag{4.11}$$

$$\min \quad (P_K^{ROW} K_{N+1}^{ROW} + P_L^{ROW} L_{N+1}^{ROW})$$

$$\text{subject to} \quad \gamma_{N+1} \left[\delta_{N+1}(K_{N+1}^{ROW})^{-\rho_i} + (1-\delta_{N+1})(L_{N+1}^{ROW})^{-\rho_{N+1}} \right]^{-\frac{1}{\rho_{N+1}}} \leq 1$$

which results in vectors of solutions

$$(Z, Z_{N+1}) = ((Z_1, \ldots, Z_N) \quad , Z_{N+1}) \quad . \tag{4.12}$$

From these solutions the output prices $(\hat{P}, \hat{P}_{N+1}) = ((\hat{P}_1, \ldots, \hat{P}_N) \quad , \hat{P}_{N+1})$

$$\hat{P} = [I-A]^{-1} Z$$
$$\tag{4.13}$$

$$\hat{P}_{N+1} = Z_{N+1}$$

are calculated[1] which assure that zero profitability hold
within each industry. At the prices (4.13) corresponding to
each of the vectors (4.9), demands for all commodities
$X(P^j)$ are calculated. Supplies are then calculated by using
output demands for all (N+1) outputs as scales of operation in each industry assuring zero excess demands for all
(N+1) outputs. The vector of supplies will be denoted by
$S(P^j)$. Thus corresponding to each of the vectors P^j in
(4.9) there is a five dimensional vector of excess demands
$f(P^j)$ which if all entries are less than or equal to zero
would imply that P^j (and the corresponding $(\hat{P}^j, \hat{P}^j_{N+1})$,
$S(P^j))$ represent a competitive equilibrium. These vectors
$f(P^j)$ are defined as

$$f_i(P^j) = X_i(P^j) - S_i(P^j) \qquad (i=N+2,\ldots,N+6) \quad . \quad (4.14)$$

The linearization assumptions are then used that

$$f\left(\sum_{j=1}^{5} w_j P^j\right) \cong \sum_{j=1}^{5} w_j f(P^j) \qquad (4.15)$$

where $\sum_{j=1}^{5} w_j = 1$; $a_j \geq -\Theta$.[2] It should be noted that tax

---

[1]The matrix $[I - A]$ need only be inverted once in
the whole calculation. Inversion of large dimensional arrays
were this necessary each time demand and supply responses
were calculated would make the calculations considerably more
expensive in terms of computer time.

[2]Θ is a parameter which limits the range of solutions which are exterior to the area of search.

revenue in this system is treated in a similar manner to other commodities, the excess demand being tax revenue collected less that distributed on the demand side of the economy. Supplies of factors of production are industry usages less endowments, demands for factors and consumptions demands, which are assumed to be zero in this application.

This enables the linear programming problem to be set up[1]

$$\min \ \varepsilon$$

subject to $\displaystyle\sum_{j=1}^{5} w_j f_i (P^j) \leq \varepsilon , \quad (i = N+2,\ldots,N+6)$ (4.16)

The objective is, through repeated use of the algorithm, to obtain a set of solutions P^j close together for which $\varepsilon \cong 0$.

To solve the problems represented by (4.16), however, a transformation of each problem is necessary. For each of the constraints, $\Theta \displaystyle\sum_{j=1}^{5} f_i (P^j)$ is added to each side and the constraint divided by $\varepsilon + M$ where M is a positive constant. This yields the constraints

$$\sum_{j=1}^{5} \frac{(w_j + \Theta)}{\varepsilon + M} f_i (P^j) \leq \frac{\varepsilon + \Theta \displaystyle\sum_{j=1}^{5} f_i (P^j)}{\varepsilon + M} \quad \text{.} \quad (i = N+2,\ldots,N+6)$$

(4.17)

---

[1]This problem is set up with constraints only on the positivity of the excess demands. From Walras Law if excess demands are all within ε , then the value of any excess supplies must also be small.

Then using the notation

$$Y_j = \frac{w_j + \Theta}{\varepsilon + M} \quad , \quad \sum_{j=1}^{5} Y_j = \frac{1 + 5\Theta}{\varepsilon + M}$$

The problem (4.16) may be rewritten as follows

$$\max \quad \sum_{j=1}^{5} Y_j$$

(4.18)

$$\text{subject to} \quad \sum_{j=1}^{5} (f_i(P^j) - C_i) Y_j \leq 1 \quad (i = N+2, \ldots, N+6)$$

where

$$C_i = \left[\Theta \sum_{j=1}^{5} f_i(P^j) - M) \right] \Big/ (1+5\Theta) \quad .$$

Problem (4.18) may be thought of as picking the weights w_j on the vectors (4.9) which minimize the largest linearized excess demand, including tax revenue. The solution to (4.18) yields a new approximate equilibrium vector $\bar{P} = \sum_{j=1}^{5} w_j \bar{P}^j$.[1] At $\bar{P}$ the excess demands are computed and if they are too large, the search process on the subsimplex is continued, another linear programming problem is constructed, and the process continued. The area of search hopefully will be systematically reduced making the assumption (4.15) more valid. Computational experience indicates $\hat{\varepsilon}$ and actual excess demands rapidly approach zero.[2,3]

---

[1] The "bars" indicate solution values for the problem (4.18)

[2] Although the uniqueness question has not been mentioned in this paper, some attention has been given to it at an empirical level. In several of the cases examined in (4.9) different initial values or paths of approach are used and the solutions are unchanged. Moreover, starting at

5. The Empirical Specification of the EEC Tax Model

To apply the computational techniques developed in
Section 4 to the model outlined earlier in the paper numerical values must be provided for all parameters of the model.
This section indicates the sources and data construction
procedures involved. Data are used for each of 5 EEC countries (France, Germany, Belgium, Holland, and Italy) for the
year 1965. The data for this year represent the most recent
data published in this form.[1] 1965 also happens to be quite
a fortunate choice in that procedures for the harmonization
of output tax systems among EEC countries had barely progres-
sed by this date. Thus the tax replacements which are considered may be interpreted as replacements of pre-harmoniza-
tion tax systems in each country by each of the other country's systems. In 1965 only one country (France) had a VAT
system and so the actual outcome of the political process
within the Commission could schematically be represented as
a common agreement on a French output tax system. The

---

solutions and moving outwards by varying degrees once again
generates the same solution.

[3]See footnote 2, p. 16.

[1]It is anticipated that data for 1970 will be available in the near future which will also include Great Brit-
ain. Recalculation with this later year data will enable
some assessment to be made of movements (increases or decreases) in the "harmonization gaps" between countries over
the five year period 1965-1970. The 1965 data are taken
from "Tableaux Entrees-Sorties 1965" published by the Statistical Office of the European Communities as No. 7 in their
Special Series in 1970.

calculations presented here enable the desirability of this
particular change to be assessed from the viewpoint of each
of the member countries.

On the production side of the economy parameters for
production processes are needed. These data come from a mix-
ture of extraneous sources and internal imposition chosen in
such a way that the numerical model is capable of replicating
the data observed for each economy as a competitive equilibrium under each country's (pre tax replacement) tax system.
The input-output data are taken from published sources for
the year 1965, each country having 55 outputs. To parameterize the functions (3.1) use is made of the data on factor
usage by industry under the prevailing tax system. Given the
observed capital labour ratio for each industry only certain
combinations of (σ_i, δ_i) are consistent with the observation
having been generated by a cost minimization procedure. The
procedure used is to fix values of σ_i and use the calculated values of δ_i for each industry. The empirical evidence on the terms σ_i for industries at the level of aggregation used here and for the countries considered is very
poor. Thus values of σ_i are taken as 1.0, and 2.0 and the
impact on the estimates assessed. The terms γ_i in (3.1)
define units of measurement for outputs and units are chosen
such that given units for factors (to be explained below) one
unit of output has a price of 1 Eur.[1] in the pre-tax replacement equilibrium.

---

[1] The data for each country are converted into "European Accounting Units" (Eur.) by use of official exchange
rates for 1965. The following are used: 1 Eur = 4 D.M. =
4.937 Fr.Fr. = 625 Lr. = 3.62 Fl. = 50 Fr.B.

To determine the factor endowments for each economy, wage bill and capital consumption data by industry published along with the input-output tables for each country are used. As the model used is static in character, endowments of factors are considered which are depleted in a single time period (factor services rather than stocks of factors). Units for factors are considered such that the physical units also correspond to prices of 1 Eur. in the pre tax replacement equilibrium. This involves a somewhat complex calculation; the net outputs (consumptions) of each commodity are known from the published data, and using the input-output matrix the gross outputs can be determined. From the gross outputs are subtracted the value of intermediate requirements in each industry. The zero profitability conditions require that at an equilibrium this valuation equals that of the factors of production. Choosing physical units for factors of production which correspond to a price of 1 Eur. in the pre tax replacement equilibria implies that this valuation is a measure of the total amount of factor inputs (capital plus labour) in each industry. This amount is then separately attributed as capital and labour using the capital labour ratio data referred to above. The capital and labour inputs used in each industry are then separately summed to obtain the endowments of capital and labour in the economy.

On the demand side of the economy one consumer is considered in each economy with one additional consumer in each case in R.O.W. Incomes are determined by sale of endowments (assumed privately owned) and the utility functions are parameterized by using the observations on the quantities consumed in the pre-tax change equilibrium. Prices net of consumer taxes (and hence gross of consumer taxes) are known and so from the demand functions (3.5) the parameters b_{ci} can be determined given values of Z_c . The terms Z_c (elasticities of substitution in the utility generating process) are varied parametrically (a procedure similar to that with σ_i); values used in each calculation are reported with each set of results.

The tax system in each country is introduced through the system of ad valorem consumer tax rates on purchases of each commodity. To obtain these tax rates data from the published input-output tables on output[1] tax payments by industry are used. These tax payments are divided by output levels to obtain "direct" tax rates and the input-output matrix used to obtain "direct and indirect" tax rates which are then used as the consumer tax rates. Table 1 lists a sample of these rates which illustrate not only the industry and countrywide differentials involved but also the discrepancies between direct and indirect tax rates.

6. Preliminary Results and Computational Experience

A sample of preliminary results from the fixed point calculations described in sections 4 and 5 is reported in Tables 2 and 3. In each case the system of consumer taxes operating in the country concerned is replaced in turn by the tax system of each of the other EEC countries. As the focus of the harmonization issues considered in this paper is on consumer taxes the equilibrium in each economy under any

---

[1]See footnote 2, p. 6.

TABLE 1

SAMPLE OF EFFECTIVE COMMODITY TAX[1] RATES BY COUNTRY FOR 1965

Commodity	FRANCE		GERMANY		ITALY		HOLLAND		BELGIUM	
	Direct	Total	Direct	Total	Direct	Total	Direct	Total	Direct	Total
1. Fisheries	0.064	0.167	0.051	0.130	0.006	0.048	0.010	0.060	0.004	0.087
2. Coal	-0.044	-0.024	0.030	0.090	0.078	0.209	-0.007	0.012	-0.037	-0.009
3. Gasoline	0.449	0.6166	0.376	0.575	0.496	0.759	0.191	0.412	0.439	0.845
4. Gas and Electricity	0.061	0.119	0.028	0.081	0.124	0.196	0.029	0.065	0.029	0.077
5. Metallic Ores	0.082	0.144	0.029	0.067	0.019	0.061	0.0	0.0	0.0	0.0
6. Iron & Steel	0.043	0.133	0.022	0.120	0.023	0.116	0.023	0.075	0.047	0.044
7. Meat	0.066	0.135	0.028	0.088	0.085	0.147	0.023	0.064	0.016	0.059
8. Dairy Produce	0.068	0.127	0.041	0.079	0.031	0.071	-0.018	-0.045	0.032	0.057

[1]See text for empirical definition of taxes and tax base.

of the alternative tax systems will be an efficient allocation, and the impact of these tax replacements upon the value
of product in the economy is of small interest. Changes in
this valuation are small (< 0.01% GNP) and reflect movements
in the composition of the output mix in the economy rather
than improvements in the efficiency of operation of the economy. Of more interest are the impacts on the utility level
of the (composite) consumer within each economy and the percentage changes involved in a sample of tax replacements are
reported in Tables 2 and 3.

There is inevitably substantial arbitrariness in the
results reported in Tables 2 and 3 as the same demand functions which are used in the model could be derived from a
process of utility maximization for any monotonic transformation of the utility functions. The estimates of proportional
changes in utility levels are not invariant to such monotonic
transformations. The estimates do, however, provide the
first evidence of which the author is aware on the relative
desirability of alternative tax systems in each country.
Indications of the importance of particular parameters for
the results are also given.

Several features of Tables 2 and 3 are worthy of some
comment. Of the country tax systems the French system seems
to represent the least desirable for the three major EEC
countries for which results are reported. Franch does better
with any other country's tax system while both Germany and
Italy are worse off with the French tax system. Thus the
agreement to harmonize output tax systems in the post 1967
period on what is basically a French system may well be the
least desirable of the alternatives considered. Of the

TABLE 2

GERMANY - IMPACT OF TAX REPLACEMENTS

	GERMAN OUTPUT TAX SYSTEM REPLACED BY TAX SYSTEM OF					REPLACEMENT BY NON DIS- TORTIONARY TAX SYSTEM
	FRANCE	HOLLAND	ITALY	BELGIUM		
CASE 1 $(\sigma_i = 1.0, Z_c = 1.0)$[1] % change in utility level	-0.41%	+0.33%	+0.02%	+0.45%		+1.41%
CASE 2 $(\sigma_i = 1.0, Z_c = 0.5)$[1] % change in utility level	-2.15%	+2.98%	+1.03%	+2.61%		+8.79%
CASE 3 $(\sigma_i = 2.0, Z_c = 1.0)$[1] % change in utility level	-0.41%	+0.32%	+0.02%	+0.45%		+1.41%
CASE 4 $(\sigma_i = 2.0, Z_c = 0.5)$[1] % change in utility level	-2.10%	+2.94%	+1.00%	+2.52%		+8.76%

[1]See text for explanation of these parameters.

TABLE 3

FRANCE – IMPACT OF TAX REPLACEMENTS

	FRENCH OUTPUT TAX SYSTEM REPLACED BY TAX SYSTEM OF				REPLACEMENT BY NON DIS-TORTIONARY TAX SYSTEM
	GERMANY	HOLLAND	ITALY	BELGIUM	
CASE 1 $(\sigma_i = 1.0, Z_c = 1.0)$[1] % change in utility level	+0.30%	+0.59%	+0.26%	+0.78%	+1.55%
CASE 2 $(\sigma_i = 1.0, Z_c = 0.5)$[1] % change in utility level	+1.95%	+5.06%	+3.06%	+5.09%	+10.98%

ITALY – IMPACT OF TAX REPLACEMENTS

	ITALIAN OUTPUT TAX SYSTEM REPLACED BY TAX SYSTEM OF				REPLACEMENT BY NON DIS-TORTIONARY TAX SYSTEM
	FRANCE	GERMANY	HOLLAND	BELGIUM	
CASE 1 $(\sigma_i = 1.0, Z_c = 1.0)$[1] % change in utility level	-0.23%	+0.12%	+0.33%	+0.65%	+1.56%
CASE 2 $(\sigma_i = 1.0, Z_c = 0.5)$[1] % change in utility level	-3.44%	-0.91%	+2.12%	+2.06%	+8.47%

[1] See text for explanation of these parameters.

individual country systems the Belgian system seems to be
the best alternative for the three major EEC countries
considered.[1]

In addition, the potential gains to each of the economies through tax replacement by a non-distortionary tax system rather than any individual country system seem substantial (between 2 and 3 times the gain from the best country
tax system). This emphasizes the point that fiscal harmonization in the EEC represents a political choice between
restricted alternatives, and that by not harmonizing on a
truly non-distortionary tax system (perhaps substantial)
welfare gains are being forgone. Tables 2 and 3 further indicate that the estimates of the potential gains from harmonization of consumer tax systems in the EEC are very sensitive to elasticities of substitution in the utility generating process but not elasticities on the production side of
the economy. The degree of sensitivity (a factor of 6 in a
movement from 1.0 to 0.5) to utility elasticities is especially striking. The sensitivity is, however, of a quantitative rather than qualitative nature and the ranking of tax
systems by size of welfare gain is preserved.

---

[1]The gains accruing to the three major EEC countries
from a harmonization on, say, a Belgian tax system seem more
substantial than estimates of gains accruing to the UK from
entry into the EEC. Miller and Spencer [11] estimate these
at 0.1% - 0.5% of welfare levels for the UK (the effect of
tariff charges excluding the impact of income transfers).
This suggests that the fiscal harmonization discussion is
perhaps of more importance for all EEC countries than tariff
policy or tariff negotiations involved during entry. Unfortunately data for the UK are not available for the 1965 period used here, but it is hoped that data for 1970 including
the UK will soon be available so that a more substantive examination of this issue can be made.

Computational experience with this model is outlined in Tables 4 and 5. For problems of this form execution time is seen to be crucially dependent on the parameters involved. The two sets of cases examined in Table 4 correspond to substantial differences in the estimates of impacts of tax replacements as reported in Table 2. It should also be emphasized that these calculations were performed on a fast machine (for instance, execution time would rise approximately by a factor of 5 if a CDC 6600 were used and perhaps 5 - 8 on an IBM 360); nevertheless for the large dimensional problems examined here execution time is not a major constraint.[1]

Table 5 reports a more detailed consideration of one of the executions reported in Table 4. The computational procedure outlined in section 4 uses several parameters which have to be specified in any execution. The indication is that some saving in execution time is possible by a careful examination of the numerical value of parameters involved. The savings involved seem to be small and probably do not justify the expense in computer time of the search procedure unless a considerable amount of execution is anticipated. The fact that the same solution is found for all parameterizations in Table 5 gives some empirical support to the contention that the competitive equilibria which are being examined in these models are unique for the particular parameterizations being considered.

---

[1]Somewhat more of a problem is the amount of core storage needed in compilation. Several (60 x 60) arrays are involved and over 30K (CDC K, the 7600 has a limit in the region of 80K) is needed. For larger dimensional problems some partitioning of arrays in addition to more efficient programming and optimization may be called for. About 5 seconds of compilation time is involved for the program used in this paper.

TABLE 4

COMPUTATIONAL EXPERIENCE ON CDC 7600

	EXECUTION TIME IN SECONDS (CDC 7600)	NUMBER OF LINEAR PROGRAMMING PROBLEMS SOLVED	DIMENSIONS AND PARAMETERS INVOLVED (SEE TEXT FOR EXPLANATION)
	GERMANY CASE 3 ($\sigma_i = 2.0$, $Z_c = 1.0$)		DIMENSIONALITY OF FULL SIMPLEX = 61
TAX SYSTEM REPLACED BY			DIMENSIONALITY OF SUBSIMPLEX = 5
FRANCE	1.52	19	DIMENSION OF CONSTRAINT MATRICES = (6 x 13) (REDUCED FROM 173 x 234)
HOLLAND	1.68	21	TOLERANCE ON EXCESS DEMANDS = 0.000000002
ITALY	1.44	18	LARGEST DEMANDS = 100,000.0
BELGIUM	1.61	20	$\tilde{\Delta} = 0.08$
	GERMANY CASE 4 ($\sigma_i = 2.0$, $Z_c = 0.5$)		$\Theta = 0.95$
TAX SYSTEM REPLACED BY			M = 0.01
FRANCE	5.65	70	PROPORTIONAL REDUCTION IN AREA OF SEARCH SHOULD INTERIOR SOLUTION BE FOUND = 2.75 (DENOTED BY PR)
HOLLAND	4.67	58	
ITALY	5.24	65	GRID SIZE (SUM) OF SUBSIMPLEX = 5
BELGIUM	4.68	58	

TABLE 5

EVIDENCE FROM VARIATIONS IN PARAMETERS OF COMPUTATIONAL PROCEDURE

GERMANY CASE 4 ($\sigma_i = 2.0$, $Z_c = 0.5$)

GERMAN TAX SYSTEM REPLACED BY FRENCH SYSTEM

$\tilde{\Delta}$	Θ	M	PR	TOLERANCE	EXECUTION TIME IN SECONDS (CDC 7600)	NUMBER OF LINEAR PROGRAMMING PROBLEMS SOLVED
.08	.95	.01	2.75	.000000002	5.65	70[1,2]
.08	.60	.01	2.75	.000000002	5.64	70[1,3]
.04	.95	.01	2.75	.000000002	5.35	66
.08	.95	.01	1.50	.000000002	5.42	67
.02	.95	.01	2.75	.000000002	5.65	70[1,2]
.01	.95	.01	2.75	.000000002	5.66	70[1,3]
.02	.95	.01	2.75	.00002	5.34	66

In all cases, the same solution is obtained.

[1] 70 Linear programming problems set as upper bound for these executions.

[2] Represents a solution within a tolerance of 0.00001.

[3] Represents a solution within a tolerance of 0.1.

7. Conclusion

This paper has reported some preliminary findings from
the application of fixed point computational procedures to
the fiscal harmonization process in the EEC. Attention has
been restricted to harmonization of output tax systems which
corresponds approximately to the current discussion of harmonization of VAT tax rates and bases in the community. A
static general equilibrium model of large dimensionality is
used which captures the distortionary features of tax systems
at a much more detailed level of aggregation than has been
possible in earlier tax replacement calculations (Shoven and
Whalley [6], Whalley [10]). Due to problems of data availability no consideration is given to the impact of tax replacements on trade flows between community members, but instead the domestic impact of tax replacements is separately
examined. Computational procedures are applied to 61 dimensional problems which use particular features of the application to reduce the execution times involved within manageable
limits.

It is found that working with 1965 data (corresponding
to the period prior to harmonization procedures coming into
operation) that the French tax system seems the least desirable tax system to be adopted by all member states, and the
Belgian is the most desirable of the country tax systems.
Relative to a non-distortionary tax system not even the Belgian tax system captures the major portion of the possible
gains. Estimates, not surprisingly, are sensitive to particular parameters but in a quantitative rather than qualitative manner (the ranking of country tax systems being preserved). Computational experience is also briefly examined
for these models.

REFERENCES

[1] Harberger, A. C., "Efficiency Effects of Taxes on In-
     come from Capital", in (ed.) - Kryzaniak, M., Effects
     of Corporation Income Tax, 1966.

[2] Johnson, H. G., "Factor Market Distortions and the
     Shape of the Transformation Curve", Econometrica, 1966,
     pp. 686-695.

[3] Scarf, H. E., "On the Computation of Equilibrium
     Prices", Ten Essays in Honor of Irving Fisher, ed.
     Fellner, et. al, pp. 207-30, New York: Wiley.

[4] Scarf, H. E., "An Example of an Algorithm for Calcula-
     ting General Equilibrium Prices", American Economic
     Review, 1969, pp. 669-677.

[5] Scarf, H. E., (with the collaboration of T. Hansen),
     The Computation of Economic Equilibria, Yale University
     Press, 1973.

[6] Shoven, J. B., and Whalley, J., "A General Equilibrium
     Calculation of the Effects of Differential Taxation of
     Income from Capital in the U.S.", Journal of Public
     Economics, 1972, pp. 281-321.

[7] Shoven, J. B., and Whalley, J., "General Equilibrium
     with Taxes: A Computational Procedure and an Existence
     Proof", Review of Economic Studies, 1973, pp. 475-489.

[8] Shoven, J. B., "A Proof of the Existence of General
     Equilibrium with Ad Valorem Commodity Taxes", Journal
     of Economic Theory, 1974.

[9] Shoven, J. B., "General Equilibrium with Taxes; Exis-
     tence, Computation, and a Capital Income Taxation Ap-
     plication", unpublished Ph.D. Thesis, Yale University,
     June 1973.

[10] Whalley, J., "A Numerical Assessment of the April 1973
      Tax Changes in the United Kingdom", unpublished Ph.D.
      Thesis, Yale University, June 1973.

[11] Miller, M. H., and Spencer, J. E., "The Static Econ-
      omic Effects of the U.K. Joining the EEC and Their
      Welfare Significance: An Attempt at Quantification in
      a General Equilibrium Framework", 1973 (mimeo).

Pricing for Congestion in Telephone Networks:

A Numerical Example

Philippe J. Deschamps

1. Introduction[*]

The study of congestion theory, with its applications to the dimensioning of telephone equipment, is far from new; it is in fact this practical problem which led to the major developments in queueing theory. The celebrated Erlang loss formula, still employed by practically every telephone administration today, dates back to 1917.

The problem of efficient pricing of telephone calls, on the other hand, has only recently attracted the attention of economists. Although a lot of work was done in the fifties on the general problem of efficient pricing for public utilities faced with periodic fluctuations of demand, (see Boiteux [1]) a telephone utility is faced with constraints of a very special nature, warranting an independent investigation.

---

[*]This paper is based on research conducted at the Center for Operations Research and Econometrics, University of Louvain, Belgium, and is to be part of the author's Ph.D. dissertation. I would like to thank foremost Professor M. Marchand, who provided me with invaluable guidance throughout. I also benefited greatly from guidance and comments by Professors J. Dreze and H. Tulkens, who raised the original question of whether fixed point algorithms could be applied to a second-best pricing problem.

I am also indebted to Professor H. Scarf for giving me the opportunity to become acquainted with fixed point algorithms during a research workshop held in Dartmouth College, and to Professor R. Saigal for introducing me to recent work on the subject.

Typing assistance at CORE and Bell Telephone Laboratories, Murray Hill, N.J. is gratefully acknowledged.

Indeed, while all public utilities producing a commodity subject to quality fluctuations resort to random rationing of demand in order to meet random fluctuations in that demand, the additional problem of determining an optimal shortage probability (or quality of service) endogenously in the pricing model takes, in our view, a special importance in this case. While a delay of one additional day in delivery for a letter, say, is probably of no significant importance to most consumers, (in other words, the price attached by society to improvements in service qualities is here likely to be low), the special vocation of a telephone service is to provide immediate conveyance of an important message from one person to another; a severe decrease in quality of service will mean that this transmission will eventually not take place at all. The following question then arises: Since providing perfect service would prove impossibly costly, what is the optimal compromise between high price and low service quality?

To the best of our knowledge, the problem of simultaneously determining a vector of optimal qualities of service, as well as prices and capacities, was first treated by M. Marchand [8] in a second best framework. The mathematical expression of the equilibrium conditions proposed presents a rather intricate level of interdependence, rendering their solution by numerical methods a virtually impossible task until a few years ago. We will be able, however, to propose a general method of solution.

Another approach is to maximize consumer surplus subject to linear constraints that impose equality between demand and supply, at a preset quality of service. This has been done by S. Littlechild [6] who gives a numerical example for a simple three-route network.

Apart from the arbitrary preselection of a standard quality of service, which may not be compatible with the value attached to it by society, such an approach does not enable one to measure the effect of the potentially large errors entailed by the linearization of the capacity constraints.

This restriction could probably be lifted in the consumer surplus model; however, the computational burden would become high, if a numerical solution can at all be found. The consumer surplus model, a partial equilibrium approach with somewhat restrictive theoretical premises, may then lose a major advantage over the more elaborate model presented here: computational simplicity.

The model we propose is very costly to implement, and fixed point methods are the only tool we know of capable of computing its solution. Becuase of this cost, these methods are practically limited to a few dozens of variables. But what telephone company would set more than a few dozen different prices? Clearly, considerations of symmetry, and the need not to confuse people with an enormous number of different rates will typically make that number much lower than its potential maximum. This constraint on the multiplicity of prices will have to be incorporated in any operational model.

Before we proceed further, it will be useful to give a short and extremely simplified introduction to telephone technology.

A network can be schematically represented as a set of exchanges, whose function is to connect a calling subscriber to the wanted number, and of junctions connecting exchanges and conveying the information.

Some exchanges will be major ones, others will be serving only a local area.

Following is the graph of a typical network without alternate routing.

⊙   major exchanges
○   regional exchanges
˙   local exchanges

Each local exchange can be directly connected with sub-scriber telephones; the connecting line is typically treated as if it were a part of the subscriber's telephone itself.

There exist different types of transmitting facilities (junctions) adapted to different distances and volumes of traffic. A junction can be treated as a set of circuits, each one of which can handle a single call at a time (multi-plexing techniques may allow several circuits to reside on the same physical medium).

An exchange, on the other hand, can be schematically represented as consisting of two types of equipment: the control 'unit' and the switching grid. The former selects a free path in the latter, and establishes the connections nec-essary to link the calling party to the wanted number or to an outgoing line.

The present model is built mainly for illustrative purposes and is suitable for networks without alternate routes
(one and only one connecting path is available between a cal-
ler and his party), with full access (a circuit has physical
access to any circuit in a consecutive junction; this assumption is realistic for crossbar and electronic exchanges) and
progressive control (a call encountering a congested equipment item is not rejected until it has been routed through
the whole subtending network). In the sequel, exchanges will
not be considered on a separate footing from junctions, and
both will be referred to as 'links', or 'equipment items'.
While this is unrealistic in view of the complexity of a modern exchange, and the associated problems in estimating congestion, it can be remedied by specifying a mixed model with
fixed shortage probabilities at switching machines, while
keeping these endogeneous for transmission facilities.[1]

2. The Computation of Optimal Rates

We shall here present a straightforward generalization
of part of Marchand's results [8] to a multi-exchange network
without alternate routing. We will then pass to the topic of
obtaining a numerical solution for the model.

I. The Model
A. Let us first make the following simplifying
    assumptions:

---

[1]On this topic and others announced in this paper,
see the author's forthcoming article on 'Second-best resource
allocation in telephone networks.

1) Calls encountering congestion are rejected (loss system);

2) The demand for telephone traffic does not depend on service qualities.

3) If a call is rejected, the consumer will repeat his attempts as long as the connection is not established.

4) The telephone company is not required to balance its budget.

B. We assume a preexisting network that has the following characteristics and parameters:

1) There are N exchanges, N^2 routes ij (where i and j are exchanges), L equipment items, or 'links' ℓ , p periods of time $r = 1,\ldots,p$;

2) Let P_ℓ be the leasing price of an additional channel (selector, circuit...) at link ℓ during periods $r = 1,\ldots,p$;

3) Let S_ℓ be the set of routes using link ℓ; E_{ij} be the set of links involved in route ij;

4) Let m_ℓ be the number of channels at link ℓ;

5) Let π_r^{ij} be the price rate of a call on route ij lasting the length of period r .

C. The equations of the model can be written as follows:

1) Let

$$s_r^{ij} = s_r^{ij} (\pi) \qquad (2.1.1)$$

be the traffic demand (in erlangs)* on route
ij at period r .

2) The quality of service on route ij at period
r :

$$q_r^{ij} = q_r^{ij} \ (\pi,m) \qquad (2.1.2)$$

can be defined as the probability that a given
attempt be successful.

3) Let us define

$$\varphi_r^{ij} = \varphi_r^{ij} \ (\pi,q) \qquad (2.1.3)$$

as the price attached by society to a marginal
improvement in q_r^{ij} .

4) Let finally

$$g_\ell = g_\ell(\pi,m) = \sum_{r=1}^{P} \sum_{i=1}^{N} \sum_{j=1}^{N} \varphi_r^{ij} \ \frac{\Delta q_r^{ij}}{\Delta m_\ell} \qquad (2.1.4)$$

be the total benefit accruing to society as a
result of the addition of one channel at link
ℓ .

Expenditure minimization assumptions for the consumer
side of the economy, competitiveness assumptions for the producer side excepting the telephone utility (implying that the

---

*Obtained by summing the durations of all offered
calls and dividing by period length.

utility buys its inputs at marginal costs), and the maximization of social utility can then be shown to imply the following optimality conditions for telephone rates and investment levels:

$$\pi_r^{ij} = - \sum_{k=1}^{N} \sum_{h=1}^{N} \varphi_r^{kh} \frac{\delta q_r^{kh}}{\delta s_r^{ij}} \qquad \begin{array}{l} i,j=1,\ldots,N \\[4pt] r=1,\ldots,p \end{array} \qquad (2.1.5)$$

$$m_\ell = \min \{ m_\ell | P_\ell \geq g_\ell(\pi,m) \} \qquad \ell=1,\ldots,L \;. \quad (2.1.6)$$

Equations (2.1.5) state that the charge for using the
system (left-hand side) should be set equal to the social
cost of the additional congestion entialed by this use of the
system (right-hand side). Equations (2.1.6) set the optimal
level of investment: equipment should be acquired as long as
the social benefit derived from adding to existing capacity
(g_ℓ) is greater than the marginal cost of equipment (P_ℓ) .

The variables π , φ and P are rates per unit of
time.

If a general way of solving numerically systems (2.1.5)
and (2.1.6) can be found, we will be able to simultaneously
compute the optimal rates and capacities at each link; optimal qualities of service are then uniquely determined by
(2.1.2).

II. The Computational Model

We will here specify explicit relationships for equations (2.1.2) and (2.1.3); the assumptions underlying these
specifications could be replaced by others, more realistic
from the point of view of technology (2.1.2) or consumer behavior (2.1.3). The present assumptions are, however,

adequate for a test of the model.

A.   The quality of service.

We assume the following to hold approximately:

$$q_r^{ij} = \prod_{\ell \in E_{ij}} q_r^{\ell}(m_\ell, s_r^{\ell}) \qquad (2.2.1)$$

where

$$s_r^{\ell} = \sum_{ij \in S_\ell} s_r^{ij} \qquad (2.2.2)$$

and

$$q(m,s) = 1 - \frac{s^m/m!}{\sum_{i=0}^{m} s^i/i!} \qquad (2.2.3)$$

So, even though an unlimited number of trials before success is assumed, we assume that the quality of service can be computed by means of the Erland loss formula (2.2.3) without too large an error.   (On the validity of this approximation, see Syski [13] pp. 592-596.)

B.   The value of consumer time.

Under some (rather severe simplifying assumptions, one of which is the assumption that the probability of success be the same for every trial, the number N_r^{ij} of attempts before success follows a geometric distribution (see [13], p. 135):

$$f(N_r^{ij}) = (q_r^{ij})^{N_r^{ij}} \qquad (2.2.4)$$

then its mathematical expectation is:

$$E(N_r^{ij}) = \frac{1-q_r^{ij}}{q_r^{ij}} \qquad (2.2.5)$$

and the quality of service becomes the inverse of the total expected number of attempts

$(q_r^{ij} = 1/E(N_r^{ij}) + 1)$.

The total expected value of the time spent trying to get connections on route ij at period r is then

$$\frac{k_r^{ij}c_r^{ij}t(r)s_r^{ij}}{q_r^{ij}} \quad ,$$

where k_r^{ij} is the average cost of an attempt,

c_r^{ij} is the average number of calls in an Erlang-hour,

$t(r)$ is the length of period r .

Summing over routes and periods and differentiating with respect to m_ℓ , $\ell = 1,\ldots,L$, yields the marginal value, in terms of consumer time, of an addition to capacity at link ℓ . This is precisely the sense of equation (2.1.4), and yields for φ :

$$\varphi_r^{ij} = \frac{k_r^{ij}c_r^{ij}t(r)s_r^{ij}}{(q_r^{ij})^2} \qquad (2.2.6)$$

which becomes our specification for the general
relationship (2.1.3).

C. Computing optimal prices and capacities.

For given m , the right-hand side of (2.1.5) is
positive, continuous and bounded. If m is allowed to vary as a function of π , however, (that
is, when the m-vector has to satisfy (2.1.6)), the
joint system of equations (2.1.5) and (2.1.6) becomes a piecewise continuous function of π ,
since m is a vector of integers.

However, since the r.h.s. of (2.1.5) remains
bounded everywhere, a very simple mathematical device can be used to extend the mapping to an upper-
semi-continuous convex correspondence. (If π^0

is a point of discontinuity, take the convex hull of
the images of the limits of points taken on the frontier of $B(\pi^0, r)$, with $B(\pi^0, r)$ a closed ball of
"sufficiently small" nonzero radium r). Since
$f(\pi)$ is defined on the whole positive orthant,
restricting the mapping to the convex hull of the
(compact) image set enables one to use the Kakutani
fixed point theorem to prove the extended mapping.

If π^* lies in the domain of continuity of the
original mapping, π^* is a solution to (2.1.5)
and (2.1.6).

The above suggests a computational procedure. It is
not too difficult to devise an efficient and globally convergent algorithm for computing a solution to (2.1.6), in
terms of m , given a vector π (this being due to the

monotonicity of g_ℓ , $\ell=1,\ldots,L$, in terms of each m_ℓ).

Fixed point algorithms such as the ones discovered by Scarf
[12], Merrill [10], or Eaves and Saigal [4] can then be used

on (2.1.5), when the optimal $m_\ell^* = m_\ell^*(m,\pi)$ are computed at

each iteration and plugged into the latter system. The algorithms in [4] and [10] have the major advantage that sufficient precision can be attained for determining whether or
not the solution lies at a point of discontinuity.

    In the next section, we describe the results of the ap-
plication of Marrill's algorithm to the model, using real-
world data. It should be noted that in the more than 20 examples that were run, no "discontinuous solution" in the
sense mentioned above was encountered.

III. An Application to Real-World Data.

    An attempt is made here to apply the model to S. Lit-
tlechild's demand and cost data [6]. This data will not be
reproduced here. The network is made up of three routes and
five "links", or equipment items, and is described by the
following matrix where an element (i,j) equals 1 when route
j uses link i :

		Routes	
links	1	2	3
1	1	0	0
2	0	1	0
3	0	0	1
4	0	1	1
5	1	1	1

Four periods are considered:

1. 6 a.m. to   6 p.m.
2. 6 p.m. to   8 p.m.
3. 8 p.m. to 12 p.m.
4. 12 p.m. to   6 a.m.

In order to convert equipment marginal costs, given in
[6] as costs per erland of capacity, in costs per circuit,
the following formula was applied:

Cost per circuit = α x cost per erland, where α
is the number of circuits necessary to process
peak-load traffic at a .97 quality of service,
divided by peak-load traffic.

S. Littlechild's capacity constraints include peak factors, so that capacity meets maximum demand within each period r, as must be the case with a commodity not subject to
quality fluctuations. Since the formal treatment of the
problem ignores quality of service, it is perfectly legiti-
mate to treat telephone service as any other commodity and to
include these peak factors in the capacity constraints. In
the present model the quality of service during period r is
the one obtaining at average, rather than peak, traffic. In
order to be able to compare the two models, therefore, the
solution to S. Littlechild's model was recomputed without
peak factors.

The social cost per erland-hour of an attempt, k_r^{ij}, was
quite arbitrarily set at 10 cents (and can be determined by
sensitivity analysis). Under the assumption that the average
costs 30 seconds of the subscriber's time, this means that

the average subscriber values an hour of him time at 12 dollars. The average cost of an attempt is then indeed

1200 cts x $\frac{30}{3600}$ = 10 cts. As in S. Littlechild's article,

the average number of calls in an erlang-hour, c_r^{ij} , is

equal to $\frac{60}{3}$ = 20 .

Table 2.3.1 presents the results obtained with S. Littlechild's model. Table 2.3.2 the results obtained with our
model. Table 2.3.3 presents the results of the application
of our model to the same cost data, but with new demand functions, introduced in [7], that include nonzero cross-
elasticities.

An interesting observation in Table 2.3.2 is that local
prices do not differ much from the prices obtained with S.
Littlechild's model. This is due to the very high quality of
service, and the quasi-linearity of the loss formula for high
traffic levels. (In fact, use of the erland loss formula in
this case is unrealistic; our conjecture, however, is that
its use does not entail a serious error for qualities of service not too far from unity.) For toll calls, prices in the
two first periods are less unequal in the nonlinear model.
This may be due to the fact that our capacity constraints
which embody variable qualities of service, are more "flexible" than fixed quality of service capacity constraints.
This is apparent on the following diagram, where A is peak-
load demand, B is off-peak demand, C is the short-run marginal cost of congestion curve in our model (for the optimal
capacities), and D the same in the linear model:

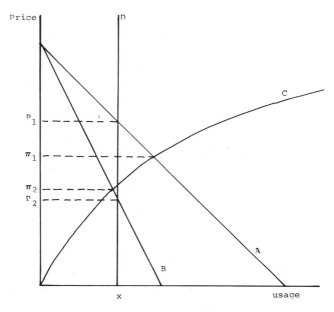

x^* is the capacity in the linear model, and P_1 and P_2 are the equilibrium prices, the prices that equalize peak and off-peak demand to capacity x^*. π_1 and π_2 are the equilibrium prices in the nonlinear model.

The ordinates of the demand functions in Table 2.3.3 are different from those obtaining in Table 2.3.2 so that the two solutions are not directly comparable. Note, however, that the gap between peak and off-peak prices has been reduced, as can be expected. Note furthermore the surprising behavior of prices on route 2, where off-peak price (period 2) is greater than peak price (period 1). This is due to the fact that equipment items 4 and 5 are shared, and carry a higher traffic during the second period.

Table 2.3.1   (Linear Model)

Route		Period	Prices*	Usage**
	1	(day)	1.10	479504
1	2	(evening)	5.83	477864
(Intra-Chicago)	3	(night)	.6	228571
	4	(after midnight)	.6	10445
	1	(day)	5.46	83
2	2	(evening)	33.63	87
(Chicago-Peoria)	3	(night)	5.4	35
	4	(after midnight)	5.4	4
	1	(day)	5.46	3168
3	2	(evening)	52.33	4805
(Chicago-New York)	3	(night)	5.4	2900
	4	(after midnight)	5.4	225

*In cents per three-minute call.

**Equivalent number of three-minute calls per hour.

Table 2.3.2   (Nonlinear Model, data as in Table 2.3.1)

Route	Period	Prices	Service Qualities	Usage
	1	1.21	.9999	478569
1	2	5.49	.9978	481169
	3	.6	1.0	228571
	4	.6	1.0	10445
	1	7.28	.9647	83
2	2	15.47	.9249	97
	3	5.42	.9996	35
	4	5.4	1.0	4
	1	7.99	.9894	3157
3	2	27.91	.6479	5344
	3	5.74	.9989	2895
	4	5.4	1.0	225

Deficit: 667 dollars/day   No. of Iterations: 2358

Final Grid Size: .0128   Significant Digits: 8

Table 2.3.3   (Nonlinear Model, Interdependent Demands)

Route	Period	Prices	Service Qualities	Usage
	1	1.84	.99968	556389
1	2	1.73	.99970	555472
	3	.6	1.0	336944
	4	.6	1.0	4109
	1	7.98	.96039	101
2	2	10.27	.98573	32
	3	5.403	.99996	32
	4	5.4	1.0	5
	1	8.98	.98809	4734
3	2	20.96	.83841	5923
	3	9.22	.98115	4841
	4	5.4	1.0	219

Deficit: 834 dollars/day

No. of Iterations: 10422

Final Grid Size: .008

Significant Digits: 5

3. Concluding Remarks

The described model aims only at presenting a very ide-alized description of a modern telephone system.   It has, however, been very useful as a computational testing ground, and lays a basis for further work intent on endowing it with an operational character.

This will form the material of a paper to come. Topics to be investigated include the addition of a budget balancing constraint [1], pricing constraints [11], metering costs [9], the specification of alternate routes and a more realistic representation of congestion in telephone networks. Finally, the traffic demands should be made dependent on the service qualities. The computational difficulties, however, are here likely to be more severe. Indeed, it is not clear that the existence of a solution can be guaranteed in this case.